Dark Star

Dark Star

A New History of the Space Shuttle

Matthew H. Hersch

The MIT Press
Cambridge, Massachusetts
London, England

The MIT Press would like to thank the anonymous peer reviewers who provided comments on drafts of this book. The generous work of academic experts is essential for establishing the authority and quality of our publications. We acknowledge with gratitude the contributions of these otherwise uncredited readers.

This book was set in Stone Serif and Stone Sans by Westchester Publishing Services. Printed and bound in the United States of America.

Library of Congress Cataloging-in-Publication Data

Names: Hersch, Matthew H., author.
Title: Dark star : a new history of the space shuttle / Matthew H. Hersch.
Description: Cambridge, Massachusetts : The MIT Press, [2023] | Includes bibliographical references and index.
Identifiers: LCCN 2023005576 (print) | LCCN 2023005577 (ebook) | ISBN 9780262546720 | ISBN 9780262376662 (epub) | ISBN 9780262376655 (pdf)
Subjects: LCSH: Space Shuttle Program (U.S.)—History. | Space shuttles—United States—History.
Classification: LCC TL789.8.U6 S6645 2023 (print) | LCC TL789.8.U6 (ebook) | DDC 629.44/10973—dc23/eng/20230215
LC record available at https://lccn.loc.gov/2023005576
LC ebook record available at https://lccn.loc.gov/2023005577

10 9 8 7 6 5 4 3 2 1

For W.E.L., who is out of this world.

Contents

Preface

This book is intended to be a somewhat late but carefully considered answer to a question put to me by one of my students in the School of Engineering and Applied Science (SEAS) at the University of Pennsylvania (UPenn) many years ago. When I presented the loss of the space shuttle orbiter Challenger and its crew shortly after launch on January 28, 1986 (and the subsequent loss of Columbia in 2003), a class member in EAS 203: "Engineering Ethics" asked me if the lesson that he should draw from these disasters was that building large machines is simply pointless. This seemed to be the conclusion of the already significant and well-reasoned scholarly canon on technological failure: that the building of complex sociotechnological systems is an endeavor almost certain to be undermined by the complexities of management or the inherent unknowability of technology itself. Much evidence supported this view, but as actual guidance for apprentice engineers trying to design machines in a more ethical and responsible fashion, it came up short. My book represents an effort not to undermine previous scholarship, but to build upon it by drawing attention once again to issues of ethical engineering design and the perilous questions of technological choice that confront every generation. It is, more than anything else, an effort to craft a more satisfying and powerful lesson for my students, who, like me, are eager to understand how they can create a better world through the things they build.

Unlike the National Aeronautics and Space Administration (NASA) projects Mercury, Gemini, and Apollo that I have written about previously, the space shuttle, or Space Transportation System (STS) as it was formally known, was a program that I had witnessed myself as an enthusiastic eight-year-old. I was too young to experience Project Apollo the way my parents and teachers had (some could not believe that so much time had passed since the Apollo

11 Moon landing, or that there could be anyone alive who could not remember it), but the shuttle would be my own Space Race. My research began in the Scarsdale Public Library in New York. Later, at Fox Meadow Elementary School, my class wrote letters to the STS-1 crew, and I saved an autopen-signed note and photograph from astronauts John Young and Robert Crippen that NASA had sent in reply. Returning to this subject as a scholar has proved as enlightening as it has been challenging. I have worked to ensure that my research does not reflect my own nostalgia about a technological program that loomed large in my own life.

The institutions that I must thank for supporting this effort have grown in size since I started researching this book. Scarsdale High School, MIT, and New York University School of Law taught me much I needed to know; UPenn, NASA, the Smithsonian Institution's National Air and Space Museum (NASM), and the History of Science Society funded my research and continue to be indispensable scholarly resources. After receiving my doctorate from UPenn's Department of History and Sociology of Science, I worked on this project while teaching and researching at four institutions. A productive year as a postdoctoral fellow working on the Aerospace History Project at the Huntington-USC Institute on California and the West gave me time to situate the shuttle's history within the history of California and the opportunity to teach in the University of Southern California's Department of History. I later refined this project as a lecturer in science, technology and society in UPenn's School of Arts and Sciences, and soon after as a lecturer in bioengineering in SEAS. Finally, my appointments as an assistant and associate professor of the history of science at Harvard University brought me into a new scholarly community that has been supportive of my research and scholarly ambitions. Additional support from the American Council of Learned Societies, Columbia University's Heyman Center for the Humanities, the Linda Hall Library, and the Institute for Advanced Study (IAS) in Princeton, New Jersey, provided me with much-needed time and space to complete this manuscript.

The people whom I wish to thank for their guidance on this project are numerous, but not too numerous to mention. I have enjoyed the continuing support of UPenn professors Ruth Schwartz Cowan, Robert Kohler, and Walter Licht, and the more recent assistance of Harvard University professors Alan Brandt, Janet Browne, Anne Harrington, Peter Galison, Evelyn Hammonds, Naomi Oreskes, Rebecca Lemov, Liz Lunbeck, Everett Mendelsohn, and Ian Miller; the Linda Hall Library's Benjamin Gross; the Heyman Center's Eileen Gilooly and Rheinhold Martin; and Myles Jackson, Joshua Horowitz, and the members of and visitors to the School of Historical Studies at the IAS during 2021 and 2022. NASM curators, administrators, educators, and fellows, including Michael Neufeld, Paul Ceruzzi, James David, Jean DeStefano, David DeVorkin, Mychalene Giampaoli, Richard Hallion, Thomas Lassman, Jennifer Levasseur, Cathleen Lewis, Jo Ann Morgan, Teasel Muir-Harmony, Valerie Neal, Allan Needell, Elizabeth Wilson, and Robert Farquhar were endless sources of knowledge and enlightening conversation. I am particularly grateful for the support that Roger Launius and Margaret Weitekamp provided throughout the preparation of this manuscript.

Archival research for this project occurred at several institutions. I am grateful to Colin Fries, Jane Odom, and Elizabeth Suckow at the NASA headquarters archives; Bill Barry, Steve Garber, and Brian Odom at the NASA History Office; the NASA Johnson Space Center Oral History Project; Elizabeth Borja at NASM for access to the Darrell C. Romick Papers, Krafft Arnold Ehricke Papers, and Robert C. Truax Collection; and Deborah Shapiro at the Smithsonian Institution archives. Amy Rupert at the Archives and Special Collections of Rensselaer Polytechnic Institute provided assistance with the George M. Low Papers, and Regina Grant and Shelley Kelly at the University of Houston–Clear Lake provided access to the Johnson Space Center History Collection. I also wish to thank Peter Collopy at the California Institute of Technology (Caltech) archives for his assistance with the Frank Malina Papers, Deborah Douglas at the MIT Museum, Tracy Grimm at the Purdue Archives and Special Collections for her assistance with the Neil A. Armstrong Papers and other collections, Shirin Khaki at the Robert Rauschenberg Foundation, Rachel Tassone and Joellen Adae at the Williams College Museum of Art, Betsey Welland of Special Collections and Archives of the J. Willard Marriott Library of the University of Utah for her assistance with the James C. Fletcher Papers, Mark Bloom for his help with the Judith Resnik Papers and other materials in the Archives and Special Collections of the University Libraries of the University of Akron, Gjovalin Nikolli for the European Space Agency Archives at the Historical Archives of the European Union, Nicole Topich for the collections of the Oskar Diethelm Library of the DeWitt Wallace Institute of Psychiatry of Weill Cornell Medical College, and the staff of the National Archives and Records Administration. I offer special thanks to William Deverall and Peter Westwick of the Aerospace History Project of the Huntington-USC Institute on California and the West, which provided me with access to the Ben Rich Papers and other unexplored archives, as well as an opportunity to move the project forward while learning more about California's aerospace history. I also spent many productive days examining Enterprise, Discovery, and Endeavour at the Intrepid Sea-Air-Space Museum, NASM's Steven F. Udvar-Hazy Center, and the California Science Center.

Portions of this work benefited from presentation at various conferences; attendees, copresenters, and commentators provided useful feedback and ample encouragement for my efforts, and I am extremely grateful to them. I am thankful for advice provided by Wayne Hale, Dennis Jenkins, John Krige, John Logsdon, Asif Siddiqi, and several anonymous reviewers who patiently waded through the proposal and draft manuscript and contributed many helpful suggestions. Friends at Harvard University, UPenn, and elsewhere, including (but certainly not limited to) Brian Daniels, Stephanie Dick, Erica Dwyer, Rachel Elder, Maria Gonzalez-Pendas, Eric Hintz, Matthew Hoffarth, Chris Jones, Amy Kaminski, Stephen Kronish, Peter Lake, Elaine LaFay, Hui Li, Amar Majmundar, Hannah Marcus, Mary Mitchell, Emily Pawley, Joanna Radin, Tina Radin, Jason Schwartz, Victor Seow, Brittany Shields, Nellwyn Thomas, Jenna Tonn, Roger Turner, Benjamin Wilson, Kristoffer Whitney, and Heather Wilmore, always encouraged me, laughed at the appropriate times, and proved the best possible sounding boards for my early work. And I would like to thank Katie Helke at the MIT Press for her early and dogged support of this manuscript. It was a pleasure to work with her on it.

Throughout, I enjoyed the support of my extended family, including my parents, siblings, nieces, nephews, uncles, cousins, and in-laws. Most of all, though, I wish to thank my best friend and partner, Whitney Laemmli, who told me to write a good book or none at all. I hope that I have.

Introduction

The word "shuttle" is spoken only once in the 1974 cult science fiction film *Dark Star*, but even so, its meaning to the audience would have been clear.[1] The term—for a spaceship that ferries cargo not to distant planets, but to better, more exciting spaceships—had been in popular use in the US for fifteen years at that point, and the National Aeronautics and Space Administration (NASA) and its contractors were hard at work building one when *Dark Star* was released. Piloted, orbital spaceplanes designed to follow the Project Apollo explorations of the Moon, NASA's space shuttles flew into space over a hundred times between 1981 and 2011 and were repositories of dreams about spaceflight's future and accommodations to the limitations of time, money, and political necessity. The space shuttle, though, was not the vehicle for which the film *Dark Star* is named or the subject of the movie's plot; rather, it was the never-seen means of supplying the titular starship with much-needed cargo. Indeed, NASA's original vision for the space shuttle was as a component of a larger space infrastructure to explore the solar system and establish a permanent human presence in space—less a vehicle to the stars than a lift to the airport.

Dark Star began as a 1970 student film at the University of Southern California (not far from the factory where NASA's first space shuttle was built); its writers, Dan O'Bannon and John Carpenter, borrowed for their film the creepy tone and deep space setting of Stanley Kubrick and Arthur C. Clarke's enigmatic 1968 art-house classic *2001: A Space Odyssey*.[2] Although dismissed by critics and theatergoers as a simple and largely unsuccessful comedic satire of the former masterpiece, *Dark Star* was something more: a meditation on failure designed for a decade of disappointment. In comparison to the stoic and capable astronauts of *2001* (who outwit the supercomputer HAL 9000 and discover intelligent life on Jupiter), *Dark Star*'s crew members are less heroic than anemic: bearded, cynical, and in at least one scene under the influence of a controlled substance. In keeping with the "stowaway" genre traditions of space science fiction, one crew member (played by O'Bannon himself) joined the multiyear mission by accident after the real astronaut (Sergeant Pinback) fell into a fuel tank. The character's actual name never rates a mention, a telling form of anonymity for a profession that, in 1970, was still producing celebrities whose names were household words.

Suggestively, the *Dark Star* itself is both a ship of discovery and an interplanetary bomber, charged with a mission that is simultaneously violent and mundane: blowing up planets with unstable orbits to make the galaxy safe for colonization. For a craft intended to support astronauts for decades, *Dark Star* is cramped, short on amenities, and filled with dubious design choices: hazardously installed wiring, oddly long elevator shafts, permanently sealed emergency hatches, and vital communications equipment housed in dangerous and inaccessible parts of the ship, like airlocks. Echoing contemporary debates about the purpose of space exploration and the appropriate level of funding for it, the hapless crew of *Dark Star* learns, moments into the film's opening scene, that funds for a desperately needed shuttle flight are not forthcoming from Congress, and the crew will have to go without lifesaving radiation shielding for the next twenty years. The crew receives this news with resignation: it is the latest disappointment for a spacecraft that exists to destroy alien worlds but that has instead been slowly killing its inhabitants. Indeed, the ship's captain, Commander Powell, died years ago in a freak electrocution—the victim of one of several mechanical issues plaguing the crew on their multiyear journey.

The future of space travel, the film posits, will be less majestic than messy: the acts of planetary destruction that serve as the ship's raison d'être are thoughtless affairs involving an orgy of mechanical switch-flipping and shouting. *Dark Star*'s crew members are doom personified, living and working on a ship that can barely sustain them, the defects of which are obvious to the audience, and to the crew as well (but too late). Indeed, the omens of the crew's eventual ruin appear early in the film, and the plot does not disappoint: a poorly designed ship built to destroy will not always do so when and where it should.

NASA's Space Transportation System (STS)—as the space shuttle was properly known—was not *Dark Star*, but the craft's development reflected *Dark*

Star's ambivalence of purpose, flirtation with routine, and hints of danger. Originally devised as a weapon of mass destruction, later envisioned as a craft to democratize space travel, and finally authorized as a vehicle designed to generate profits, the shuttle was confused in its design, construction, and use—an example of how much technology had outstripped its purpose in the later years of the first Space Age. Astronauts alienated from society, without a sense of purpose or even a basic funding mandate, riding atop a bomb to create a better future: the crew of *Dark Star* and the astronauts of NASA's space shuttle shared more than a little in common.[3]

The Shuttle and the Dream of Space Travel

With American human spaceflight acutely dependent on congressional funding, press enthusiasm and political support have always proved to be as influential in structuring America's haphazard agenda in space as military and scientific need.[4] American human spaceflight, as Walter McDougall chronicled first in 1985 in *. . . the Heavens and the Earth: A Political History of the Space Age*, did not emerge in a moment of calculated calm, but rather from a "media riot" following the Soviet launch of the first artificial Earth satellite, Sputnik 1, in 1957.[5] Under a reluctant Republican president, Dwight D. Eisenhower, Congress chartered NASA, and the new agency recruited and publicized a force of military pilots to challenge Soviet civilian space achievements. Later, the Democratic president John F. Kennedy seized upon public impatience and the perceived American lag in space technology and fixed upon a new goal: a voyage by an American to the Moon, which would guarantee the nation's global standing and, thanks to television, make heroes of the aviators who achieved it.

Much more than superpower competition motivated these voyages into space. As later historians recounted, accompanying the construction of rockets and spacecraft was a new culture of exploration that imagined the extension of humanity into the cosmos.[6] Many space enthusiasts both inside and outside NASA during Apollo's voyages to the Moon anticipated the imminent transformation of humanity into a transplanetary (or even transstellar) species. Some even saw space travel as the answer to the ills and dangers that had plagued Earth for millennia: war, environmental degradation, overpopulation, or asteroid impact.[7] After exploring the Moon, futurists working with NASA's Ames Research Center near San Francisco hoped that humans would settle by the thousands in metal-and-glass colonies trailing Earth in orbit around the Sun, as Europeans had settled in the forests of New England 400 years earlier.[8] (That the word "colony" might evoke a millennium of conquest for much of Earth's population did not always occur to some of the Euro-Americans who articulated these ideas, but others employed the term "space settlement" instead.)[9] Eventually, space settlements would be supplemented by planetary bases, cities, and civilizations built on Mars or some heavenly body not yet explored. In these new paradises, prejudice, superstition, and violence would find no quarter: "Space is the place," as the Afrofuturist musician Sun Ra (born Henry Poole Blunt) explained in the 1974 science fiction film of the same name.[10] For him, space settlement meant liberation and escape—involuntary, if necessary—of people of African descent from a planet that had never treated them well.

To spaceflight's most ardent admirers, like US army major-general John Medaris, rockets yielded not merely strategic superiority, but "a new understanding of man's relationship with the infinity of Divine Creation.'"[11] Some members of the public were skeptical—like the schoolchildren who wrote to the Nazi émigré and rocket pioneer Wernher von Braun expressing their fear that his rockets would harpoon angels on their ascent[12]—but others saw spaceflight for what it could be: a chance to hedge their bets on Earth's survival, start over, and become something more than merely human. On distant planets, some radicals argued, a great experiment in human evolution would begin: humanity would replace quaint and archaic notions of the divine with something more durable, planted on the sounder footing of our infinite capacity for invention. "We are as gods," Stuart Brand wrote in the original 1968 edition of the pro-space exploration *Whole Earth Catalog*, "and might as well get good at it."[13]

But despite McDougall's claims otherwise, the space shuttle was never part of this "liberal agenda" that supposedly inspired Project Apollo.[14] Out of the White House by 1969, Democrats were in no position to determine national space policy, and their Republican adversaries had different priorities in mind. Republicans had long been wary of civilian spaceflight expenditures but eagerly sought a Potemkin space program that would suggest continued American competitiveness while limiting fiscal commitments and political exposure. For them, spaceflight offered solutions to far more mundane military, economic, and political problems like geopolitical reconnaissance and defense-sector employment.[15] Politicians like Eisenhower had been skeptical of civilian spaceflight but lavished funding on its military sibling; his former vice president, Richard Nixon, approved the space shuttle a decade later when he was president, in part because it shifted civilian space funding to the military.

Although utopian enthusiasm may have motivated some space futurists at the time, the real engines that drove the shuttle's development were war, commerce, and politics. The concept of an orbital spaceplane first emerged in Austria before World War II; Germany, the Soviet Union, and the US later sought it as a bombing aircraft with intercontinental reach. Its inventors also championed it as a technology to make space travel more routine, but it would likely cost a fortune to develop. By 1969, this concept had evolved into a $12 billion concept for a reusable rocket booster a detachable winged glider. The latter would orbit the Earth between 100 and 300 miles above the ground, and both would return to the ground and land like airplanes on conventional runways. Although never leaving low Earth orbit, the shuttle orbiters (NASA and private entities might purchase dozens)[16] might usher in a new age of space travel; astronauts arriving in Earth orbit on space shuttles might even transfer to space stations and interplanetary rockets that would take them to Mars and beyond.

This last part was pure fantasy; throughout its development, the shuttle found the greatest support for less ambitious reasons: NASA hoped to retain its funding after Project Apollo, the Air Force and secret National Reconnaissance Office hoped to launch spy satellites, and private industry hoped to use the shuttle for commercial space launch operations and tourism. Indeed, the shuttles' comfort and reusability promised to provide reasonably easy access to Earth orbit for paying passengers. Changes "in modes of flight and re-entry" associated with the shuttle, Nixon declared in 1972, would "make the ride safer, and less demanding for the passengers, so that men and women with work to do in space can 'commute' aloft, without having to spend years in training for the skills and rigors of old-style space flight."[17] Commuter transport, vacation hotspot, colonial outpost, laboratory, factory, fortress: a spacecraft that could sail into orbit so easily and cheaply promised to revolutionize, democratize, and economize space travel,[18] making humans a true spacefaring people.[19]

Sweeping language aside, though, this was a space enterprise of profit, power, and politics, not idealism. Clad in a kind of muscular nationalism, the space policy of the so-called New Right sought taxpayer funds to pay private companies to develop expensive new space weapons in the hope that these subsidies would be recouped through commercial spaceflight activities using the same technologies.[20] Instead of exploring space, NASA lavished funds on defense contractors to develop space vehicles; then, once spaceflight became cheap, NASA would convert itself into a profit-making enterprise, competing with private companies for the international satellite launch business and filling America's coffers instead of emptying them. Only later did conservatives realize that their supposedly business-friendly space policy flirted with state control over industry: by 1986, NASA's space shuttle was undercutting competing private launch providers and restructuring American spaceflight along the lines of its Soviet counterpart.

Even less noble, as John Logdson has noted, the shuttle ultimately won Nixon's support in 1972 for the most mundane and self-serving of reasons: he saw personal political benefits in its construction.[21] The space shuttle provided what the Office of Management and Budget (OMB) deputy director Caspar Weinberger regarded as a technological talisman to assuage skittish voters fearful of America's decline,[22] while other influential voices claimed that it would enrich a major Nixon campaign donor and sustain aerospace employment in California, a swing state critical to his reelection campaign.[23] Addled by reelection woes, Nixon assented verbally to NASA's plan, not quite certain of what he had agreed to and unconvinced that the plan would produce an economical space vehicle. But, enthusiastic about the toy model of the shuttle that NASA lent him for the press conference announcing it, Nixon stole it.

The space shuttle was daring; it was messianic; and it failed. The system of Earth-orbiting and interplanetary spaceships favored by radicals and intended to provide a destination for shuttle missions proved more expensive than Nixon would support,[24] leaving NASA to develop a space shuttle with nowhere to go. And the vehicle that NASA's contractors ultimately produced, while meeting Nixon's budgetary constraints, was fiendishly complex and steeped in design compromises and lost opportunities. NASA's promise that the military services and intelligence community could borrow the craft to launch and recover spy satellites ensured its approval but dictated its size and shape. Initial plans to make both the booster and the orbiter fully reusable proved too difficult and expensive; the half-price shuttle, eventually built for $6 billion, was a collection of reusable, semireusable, and disposable parts. The thermal protection system (TPS) for the orbiters—thousands of ceramic tiles intended to absorb frictional heating when the orbiters slammed into the atmosphere at the conclusion of their mission—proved finicky and delicate.[25] On the shuttle orbiter Columbia's first flight in 1981, nine years after its approval, troubling problems arose that proved fatal on later missions, and the shuttle's crew members eventually found that flying the vehicles required equal or greater acceptance of risk compared to what previous spacecraft had demanded. Shorn of many of its features, yet still hideously expensive to build and operate, the vehicle was eventually constructed enjoyed the unqualified support of no major constituency.

By the time the shuttle fleet finished its thirty-year flying career in 2011 with 135 missions, the program had flown (and killed) more people than any spacecraft before or since. Despite two accidents resulting in the loss of fourteen crew members and two shuttle orbiters in 1986 and 2003, the STS continued to fly, orbiting satellites and the Spacelab and SPACEHAB flying laboratories and helping to assemble the International Space Station (ISS). By the end of its operational lifetime, the shuttle fleet had even become the indispensable American technical infrastructure that its advocates had wanted: the only American vehicle left that could send humans into space and bring them back. These benefits, though, were accidental rather than intended: by every measure, the shuttle had fallen far short of even the modest hopes that had surrounded it. And the shuttle remained flying only because every effort to replace it with a better-winged, reusable craft also failed.

No matter how well intentioned the shuttle's evaluation by its advocates may be—that it ultimately achieved a great deal despite its inauspicious beginnings, flew hundreds of people into space, performed remarkable science, and helped enable critical programs of space exploration[26]—the shuttle did not (and indeed could not) have become the vehicle that many of its earliest enthusiasts had hoped it would be. This failure was not due to any unsolved scientific or technological puzzle, lapse in management or maintenance, or act of malfeasance by those in charge of operating the shuttle. The vehicle's flaws ultimately lay in the shuttle's design and its fundamental lack of purpose—both ultimately aggravated by NASA's fear of organizational irrelevance in the post-Apollo era. Ample evidence suggests that NASA and its contractors recognized the impossibility of their ambitions early on: the shuttle's failure occurred principally because those who designed the vehicle never intended it to succeed as anything more than a flawed placeholder for a future that they could not discern and a set of needs that they could not define.

Even as a Potemkin program, the shuttle was a failure. In the decades since engineers had first imagined it, changing technology had rendered the space shuttle obsolete, an example of what David Edgerton (2007) called humanity's obsession with the "futurology of the past."[27] The choice of this futuristic but flawed craft retarded rather than advanced the technology of space exploration, as NASA first rejected incremental improvements to Project Apollo's proven technologies in favor of a long-desired, futuristic architecture, and then found itself unable to move past the shuttle after its arrival.

Writing about the Space Shuttle

In the literature focusing on invention and innovation, space exploration has sometimes been dismissed (not entirely unfairly) as a Cold War military display technology: a government-funded distraction, removed from the normal dynamics of private innovation and receiving excessive public attention given its actual value to society.[28] (That this public attention may have been artificially constructed to smooth legislative support for big-budget space programs is the subject of noteworthy new scholarship.)[29] This interpretation may have something to do with the fact that innovation models focused on private-sector innovation do not reflect the degree to which government investment (particularly military investment)[30] has played a massive role in the creation of virtually all technological infrastructures.[31] Space travel is unusual not because it represents any kind of fundamental break with the ways in which new things come into being, but because it offers a particularly clear example of a pervasive relationship. With nine out of every ten dollars appropriated to NASA used to pay domestic contractors,[32] the American space program of the 1960s made plain the fact that many of the defining technologies of the twentieth century became ubiquitous only because the US government decided that they should be. For example, rockets were fabricated from tanks, pipes, and pumps, but also transistors. By making rockets a mass production item, the US government made semiconductors a commodity and created the modern computer industry by training the tens of thousands of experts required to use them.[33]

The last forty years have seen ample academic and popular scholarship on space history, beginning with the earliest American human space programs: Mercury, Gemini, and Apollo.[34] Historical attention has recently shifted from the well-documented Project Apollo to programs of the 1970s, 1980s, and

1990s.[35] The space shuttle has largely escaped comprehensive scholarly study, although a number of works endeavored to keep track of operational milestones of the program or to focus on specific payloads or missions.[36] Some of these works take the form of institutional histories or data books published by NASA or prepared under contract throughout the shuttle's operational life.[37] Noteworthy volumes include T. A. Heppenheimer's two-volume study of the origin and development of the space shuttle[38] and the former shuttle program manager Wayne Hale's *Wings in Orbit: Scientific and Engineering Legacies of the Space Shuttle, 1971–2010*,[39] as well as the compendium of critical documents in NASA's "Exploring the Unknown" series. This multivolume anthology mined NASA's historical archives for critical memoranda in the development of the space shuttle and published them with ample annotation by leading scholars.[40] These works are supplemented by operational histories privately written by former shuttle engineers like Dennis Jenkins, whose flight-by-flight account of shuttle operations, *Space Shuttle: The History of the National Space Transportation System*, is the gold standard of reference works in the field and has been updated to tell a complete technical history of the program.[41]

The difficulties of accounting fully for an ongoing space program previously made the preparation of historical monographs on the shuttle program difficult.[42] Memoirs were among the earliest interpretative works on the shuttle; the first of these came from NASA veterans of the 1960s,[43] like former chief astronaut Donald "Deke" Slayton (one of NASA's "Original Seven" astronauts of 1959 and the manager who led shuttle development in the 1970s) and a number of shuttle astronauts once their flying careers ended.[44] Some of the accounts emphasized the novelty of the shuttle experience, particularly that of scientist–crew members whose activities in space, including spacewalks: extravehicular activity (EVA) often captured more popular attention than the traditional piloting that the shuttle also required.[45] Other memoirs were more self-conscious efforts to harken back to Tom Wolfe's 1979 account of NASA's early years, *The Right Stuff*,[46] suggesting that the age of the heroic astronaut had passed despite some shuttle astronauts' best efforts to reclaim it.[47] And a small but growing body of popular literature has begun to examine issues of professional dissatisfaction in the shuttle astronaut corps.[48]

Historians have also attempted to provide accounts of the shuttle program by examining facets of its development and operations or by annotating the first-person recollections of astronauts and other historical actors.[49]

The Hubble Space Telescope (HST), launched in 1990,[50] was an early subject of study, as was NASA's management of the shuttle,[51] the experience of shuttle astronauts aboard the Russian Mir space station,[52] the development of the shuttle's unique winged flight mode,[53] and the torturous process that led to the shuttle's approval in 1972.[54] Women astronauts in the space shuttle program received attention from Bettyann Kevles's *Almost Heaven: The Story of Women in Space*[55] and Amy Foster's subsequent, thorough treatise *Integrating Women into the Astronaut Corps: Politics and Logistics, 1972–2004.*[56] The space shuttle's national security role, though challenging to research given the paucity of unclassified primary sources, has been examined as well.[57]

Although born during the Cold War superpower competition and the domestic political exigencies that sustained it,[58] the space shuttle outlived both the Soviet Union and the reasons for its own existence.[59] Valerie Neal, curator emerita of post-Apollo spaceflight at the Smithsonian Institution's National Air and Space Museum (NASM) published one of the first cultural histories of the program, *Spaceflight in the Shuttle Era and Beyond: Redefining Humanity's Purpose in Space,* which drew upon methodologies in American and visual studies to account for the program's distinctive iconography and historical legacy (including the eventual use of several shuttle orbiters in museum displays).[60] Work by curator and space history department chair Margaret A. Weitekamp has similarly explored the rich body of toys and popular entertainment about the space shuttle program.[61]

The present study is neither an account of the cultural meaning of post-Apollo spaceflight (an effort undertaken by other historians),[62] nor an encyclopedic history of the development and operation of the shuttle orbiters Columbia, Challenger, Discovery, Atlantic, and Endeavour (a feat ably accomplished by other scholars),[63] nor is it an ethnography of NASA's workforce (another area of previous scholarship).[64] Rather, this book is an effort to answer a single fundamental question about the space shuttle: why did it fail to make spaceflight cheap, safe, and routine? The concept of failure—technological and otherwise—has been a subject of substantial study for historians of technology, and of American history and culture more generally. Mentions of Challenger have occasionally appeared in these works as a kind of macabre poster child of failure,[65] but otherwise it has escaped proper study.

Failures, though, are not always completely negative: Scott Sandage, in *Born Losers,* writes that in the nineteenth century, Americans defined failure as "an incident, not an identity," attributable more to bad luck and an abundance of ambition than a deficiency of will or intelligence.[66] Building on this idea, Edward Jones-Imhotep writes of technological failures (particularly those involving conflicts with nature) as generative forces in the building of national identity.[67] The space shuttle, in the smattering of works that have emerged about it, is often described in this way: as a brilliant but overambitious first draft to a new and better Space Age challenged by the forces of physics. When the space shuttle began flying in 1981, the classic accounts continue, too many people expected too much of it until the shuttle, overtaxed, reached its breaking point. This is a narrative that many chroniclers of the shuttle era have previously accepted,[68] even the author of the present study, in prior work.[69] Only occasionally have scholars chosen to question this interpretation, wondering instead if the classic account is too simple.[70]

A few critical assessments by scholars emerged soon after the shuttle's early flights. Alex Roland's "Triumph or Turkey?" published in *Discover* in November 1985, presciently suggested the imminent collapse of a space infrastructure that did not appear to be functioning in the ways that its designers promised it would.[71] Challenger's loss in 1986 also attracted a flurry of scholarly interest; among the most influential scholarly books on the subject was the sociologist Diane Vaughan's 1996 *The Challenger Launch Decision: Risky Technology, Culture, and Deviance at NASA*.[72] A defining work in both space studies and the sociology of organizations, it argued that engineers unconsciously allowed known problems with the shuttle to fester due to their irrational faith in technology and NASA's broken safety culture.[73] A sensation when it appeared, *Challenger Launch Decision* remains the foremost work on the disaster and the shuttle program itself. (Similar volumes followed the loss of Columbia during reentry in 2003.)[74] Retrospectives appeared over the course of the shuttle program, attempting to contextualize historically the shuttle's economics and ascertain how badly it had fallen short of its promises.[75]

This book is both a continuation of the work of historians like Harry Collins, Roger Launius, John Logsdon, Michael Neufeld, Trevor Pinch, and Alex Roland and a departure from it. Difficult as it may be to say, the shuttle was not the brilliant failure that many its commenters claimed it was, but rather a budget-conscious attempt to solve a very difficult problem, oversold in its capabilities and value to the nation. In its thirty-year flight history, the space shuttle demonstrated impressive capabilities, but many of these achievements were accidental rather than intended, with some invented after the shuttle had already failed at its principal mission. The shuttle program also suffered two tragic losses of flight vehicles and crews, attributable to components that malfunctioned in simple, predicted, and inevitable ways connected directly to their design.

The shuttle did not fail because it was large, complex, or expensive; if anything, the shuttle was too small, and the funds spent on it insufficient. Likewise, the shuttle story is not one of a hapless government bureaucracy badly performing management and maintenance tasks better left to the private sector—NASA was the best institution of its kind in the world at the time of the shuttle's creation, and its preferred design for the shuttle was vastly superior to the version eventually constructed. The shuttle was simply built the wrong way, at the wrong time, and for the wrong reasons. Its deficiencies were obvious to those who designed and flew it, and were left unremedied not because of callousness or carelessness, but because they could not be fixed: the shuttle's vulnerabilities, like those of the *Dark Star,* were imprinted into its design at inception. Indeed, in 1982, the Santa Monica–based RAND Corporation (the former research subsidiary of the Douglas Aircraft Company) predicted that from one to three shuttle orbiters would be lost during the life of the program, a conclusion of depressing accuracy.[76] NASA always expected to receive funds to build a replacement vehicle that remedied the shuttle's problems; their obviousness, in fact, was described as a benefit, making the shuttle's replacement a near-certainty. *The shuttle failed because it was designed to fail.*

The management of risk[77] and the obsolescence of technologies,[78] both accidental and otherwise, have been perennial subjects of study for historians of technology, social critics, and activists. Arwen Mohun writes that during the twentieth century, the "vernacular risk culture" of American life became mediated by probabilistic assessments as the risks that Americans confronted increasingly stemmed from their interaction with large machines.[79] As personal and theoretical knowledge of risk began to clash, the question of appropriate risk became one of "safe enough" rather than safe, and the question of who would define it became a political one of great complexity.[80] In his 1965 book *Unsafe at Any Speed: The Designed-In Dangers of the American Automobile,* the consumer safety advocate Ralph Nader placed the blame for the modern American automobile's unimpressive safety record on the glaring, known flaws in the design of highly touted models like the Chevrolet Corvair. Rather than addressing these flaws, Nader wrote, manufacturers relied upon users to compensate for them through aftermarket improvements, or for government regulation to ameliorate the risks—not by mandating design changes, but by modifying users' behavior.[81] ("Corvair" was even the name assigned to a fictional shuttle orbiter in a 1994 episode of the animated television series *The Simpsons*.[82]).

The continued use of seemingly obsolete technologies has received ample study among historians of technology like David Edgerton, much of it devoted to the surprising durability of older inventions and the dangers of futurism."[83] Lee Vinsel and Andrew Russell have focused attention not only history of technology's obsession with stories of invention, but with the acute problems associated with maintenance of older technologies.[84] The fact that the shuttle was, by the end of its operational life, a nearly forty-year-old technology was not the source of its problems: rather, it was the shuttle's embrace of novelty in 1972 that inspired designers to abandon older and more reliable technologies when creating it. The shuttle's obsolescence, furthermore, was engineered as a cost accommodation and possibly a spur to its continued development.

"Planned obsolescence"—the purposeful introduction of technologies with limited life spans—first emerged in connection with the General Motors Company's introduction of the "model year" concept in 1924: a series of minor, mostly cosmetic annual changes to automobiles intended to spur new sales. Vance Packard, in his 1960 book *The Waste Makers*, expounded on this idea, criticizing the efforts by manufacturers, retailers, and marketers to reduce the life span of new products to increase the sales of the replacements.[85] The shuttle was not a consumer product with an intentionally short shelf life, but it was built with the expectation of rapid replacement, and its deficiencies were obvious to those who designed and flew it. Rather than providing a case study on the perils of flawed management cultures or the unknowability of complex technologies, the space shuttle instead offers lessons on the importance of ethical design, the dangers of planned obsolescence, and the critical role of human agency in the management of risk.

Organization of the Study

The origins of the space shuttle lie in a series of earlier, similar-winged, rocket-powered airplanes, each of which demonstrated its impracticality before it flew. Chapter 1 of this book, "The Silver Bird Comes to America, 1941–1963," explores the controversies later found in the space shuttle in earlier debates about the capabilities of earlier rocket plane designs. A goal of aeronautical engineers stretching back to the late 1920s, the concept of an airplane that could fly into space captivated German (and, later, American) designers throughout the 1930s, 1940s, and 1950s, despite a growing body of evidence for the vehicle's technical limitations and lack of utility. Enchanted by the potential capabilities of such a craft and undaunted by its design challenges, the US air force and its contractors lobbied extensively in the 1950s for a military space program built around the technology, attempting to justify it with an amorphous, changing, and often fraudulent repertoire of supposed benefits.

The subject of intense Air Force fascination in the 1950s but ultimately rejected in 1963 in favor of space capsules, the spaceplane found new life in 1968 as a civilian program to follow Project Apollo's successful lunar explorations of 1969 through 1972. Considered untenable only five years earlier, the spaceplane promised to return space flying to the aviation roots prized by the military flight test community that formed the core of NASA's civilian human spaceflight program. Chapter 2, "Star Clipper and Other Fantasies, 1963–1969," describes the growing enthusiasm in civilian and government circles for a spaceplane to fulfill NASA's ambitious plans for post-Apollo interplanetary exploration. This enthusiasm manifested both in NASA symposia on reusable space vehicles and a flurry of contractor proposals, including the ambitious Lockheed design referenced in the title of the chapter. Central to these schemes were conscious efforts to combine, in the minds of policy analysts, legislators, and the public, the separate concepts of piloted flight, winged flight, and reusability. By arguing that only a winged space vehicle flown by human pilots would be reusable and inexpensive to fly, the shuttle's advocates created a powerful cudgel with which to defeat competing proposals for alternative space vehicles offering better reliability and economy.

Designing a spaceplane that was palatable to the constituencies supporting it proved more challenging than first expected. Chapter 3: "The Winged Gospel and National Security, 1969–1974," describes the tortuous process by which NASA abandoned its ambitious early plans for the space shuttle in favor of an airplanelike vehicle of more use to NASA's influential partner, the Air Force. While both institutions wanted a winged spacecraft, NASA sought a leaner spaceplane to accomplish its mission of interplanetary exploration, while the Air Force sought a bigger vehicle whose missions kept it closer to Earth. Critical to Air Force plans was a shuttle with a massive payload capacity and wings large enough to give it ample cross-range: the ability to maneuver through reentry into Earth's atmosphere and land on a runway near its launch pad after depositing spy satellites in low-Earth orbit. Such a craft, though, would be heavier, less reusable, and more vulnerable, pitting NASA's exploration goals against competing desires to militarize the space environment. Drawing on Joseph Corn's seminal work on aviation as a secular religion in the US[86] and early assessments of the shuttle design process,[87] this chapter examines the shuttle's relationship to the metaphor of terrestrial airplane travel, expressed through both NASA's enthusiasm for the shuttle's potential utility as a transportation technology and its insistence that the craft have wings, a second, more pernicious "winged gospel." So eager were the shuttle's advocates to build an airplanelike vehicle that NASA and the Air Force sacrificed reusability as a system goal, undermining the principal justification for the shuttle's construction.

At one time intended to be the ambitious linchpin of an entirely new space exploration infrastructure, the space shuttle became, by the time of its first flight in 1981, a placeholder craft, operating in isolation, that NASA intended to replace with a safer and more capable space vehicle that never arrived. The shuttle's design issues were an open secret throughout the program's early flight years; chapter 4, "Tickling the Dragon, 1974–1986," describes the "go-go" years of the space shuttle program from its initial Approach and Landing Tests (ALT) to the mid-1980s, when NASA pushed the operational limits of the shuttle's design in a desperate and ultimately unsuccessful attempt to build public confidence in the vehicle. Taking its title from the Manhattan Project scientist Richard Feynman's nickname for the hazardous and ultimately fatal criticality experiments conducted on atomic bomb cores at the Los Alamos Laboratory in the 1940s, the chapter examines the deadly yet inescapable design issues that eventually overwhelmed the shuttle program. Modeled on the types of spacecraft that NASA's pilot-astronauts had always dreamed of flying but built imperfectly, the shuttle required constant repair and modification to stay flight-worthy, as its designers had always expected. Yet while NASA's management decisions during this period attracted significance criticism from later commentators, they likely did little to alter the fundamental vulnerabilities of the STS, which were well known to those building and operating it.

Chapter 5, "A History Rooted in Accident, 1986–2011," examines the losses of the shuttle orbiters Challenger in 1986 and Columbia in 2003, which independent investigators blamed on management failures but were instead tied directly to NASA's fateful decisions on the configuration of the shuttle in

1972, as described in chapter 3. Rather than excusing the loss of Challenger and Columbia as symptoms of egregious management failures[88] or the kinds of incomprehensible crises that supposedly affect all complex technological systems,[89] this chapter identifies the disasters as an entirely predictable—and predicted—consequence of a well-understood defects in the shuttle's design. The failure of a gasket on the Challenger's solid rocket boosters (SRBs) and the breach of Columbia's TPS had occurred before and was known to be potentially deadly; there was simply no solution to either problem that would not result in grounding the shuttle pending a redesign. NASA's failure to address these known vulnerabilities was less a management breakdown than a choice made to continue the US human space program, even if it could not conduct it safely with the funds provided. Inverting the title of chapter 5 to echo the 1986 *Report of the Presidential Commission on the Space Shuttle Challenger Accident* ("An Accident Rooted in History"[90]) alludes to the decades in which shuttle operated as a flawed vehicle, troubled by a history of costly design mistakes and resultant catastrophic failures.

So dependent had the US become on the space shuttle by 2003 that even following a second catastrophic shuttle accident, NASA was unwilling to halt its use. Chapter 6, "The Quest for Alternatives, 1972–2011," steps back from the operational history of the space shuttle to examine three efforts to imagine substitute architectures. After considering these alternatives, this chapter explores the parallel Soviet experience with spaceplanes. Having made similar investments in expendable launch vehicles and spacecraft technology in the 1960s, the Soviet Union balked at first at the Americans' obsessive embrace of the space shuttle in 1972 and, finding no public, peaceful reason to build one, assumed that the US had found a secret, belligerent one. In attempting a decade later to partially replicate space shuttle technology with its Buran vehicle, the Soviet government soon realized that in a political environment characterized by autocracy, a space vehicle designed to galvanize electoral support held no value. Even more, the single, unpiloted flight of Buran punctured the rationale that sustained the shuttle as a human spaceflight program.

After the concept of the spaceplane proved untenable in two countries, NASA throughout the 1980s, 1990s, and 2000s sought to replace the shuttle with a succession of newer spaceplanes, supposedly free of the shuttle's glaring faults. This chapter offers a story of failed innovation as NASA struggled to cling to the winged flight mode on which it had bet its future. An epilogue concludes the book, assessing the shuttle's risks, reflecting on its final years, and speculating on the future of American human spaceflight.

Why the Shuttle?

A vehicle of staggering and unprecedented complexity, comprised of 2.5 million individual moving parts, the space shuttle was a demonstration of American technical capability that challenged a generation of aerospace engineers to create new solutions to vexing problems in structures, materials, propulsion, and guidance. Over three decades, this amalgam of parts ultimately rocketed into space 134 times, carrying 848 crew members with it. An examination of many of the technical decisions that produced this vehicle is vital to understanding why it was built the way it was and how decisions about the shuttle's design limited the flexibility of engineers wrestling with the system's many compromises and vulnerabilities.

The fact that engineers believed they needed to make this troublesome vehicle work owed much to their conviction that the space shuttle was an inevitable technology, the emergence of which was necessary not so much to preserve the US lead in space exploration, but to advance the craft of aerospace engineering. As the historian Thomas Hughes articulated in his discussion of "technological momentum," humans may construct technologies, but thereafter *they* construct *us*, limiting our vision and constraining our choices.[91] Ultimately, the story of the space shuttle is less the story of a grand infrastructure that fell short (as described in countless scholarly and popular works) than a temporary, flawed solution to a problem that did not exist—and thus could never be solved.

1 The Silver Bird Comes to America, 1941–1963

The craft would be fast. It would fly high. And its target would be New York City, which lay untouched by World War II's strategic bombing campaigns. The unsuccessful entry into Nazi Germany's secret 1941 "Amerika Bomber" competition proposed by the Austrian engineer Eugen Sänger and the German mathematician Irene Bredt was not a multiengine, propeller-powered airplane like those offered by German airplane-makers Heinkel, Junkers, and Messerschmitt, but a chubby metal craft with a wide, flat bottom, stubby wings, and a liquid-fueled rocket engine. (Sänger was vague about the fuel, but kerosene and alcohol were likely candidates, with liquid oxygen as the oxidizer.)[1] After accelerating to takeoff speed on a rocket sled in Germany, the Silbervogel (silver bird) would climb under its own power for five minutes, eventually coasting to an altitude of 162 miles above the Earth's surface. It would then descend and climb again, unpowered, in eight skips across the atmosphere until it reached New York City, where it would drop a 660-pound bomb before gliding to a landing field in imperial Japan, halfway around the world from its launch point. Flying alone, it could be an antiship weapon, or perhaps a tool of international assassination; in waves, Sänger promised, squadrons of such craft could obliterate cities anywhere on Earth within days.[2]

Silbervogel was not the space shuttle later built by the National Aeronautics and Space Administration (NASA), but it was, perhaps, the closest thing to it that engineers had devised by 1945, and the closest they would come to it for another twenty-five years. Although scratching the edge of space, the Silbervogel could not orbit the Earth as NASA's later space shuttle would: the rocket sled that launched Silbervogel and its own internal rocket engine could lift it beyond the Earth's atmosphere but would be incapable of blasting the craft to the speed required for it to sustain an orbit around the Earth, or decelerate it enough to safely return to the ground if it did.[3] Instead, the

Silbervogel's flight would be a series of suborbital hops into space, with the craft's metal skin protecting it from the extreme frictional heating that the craft received each time that it bounced off of Earth's atmosphere.[4] Critically, the audacious design promised to perform the key feat with which the shuttle would later be credited: controlled, winged flight into and out of the atmosphere.

Reflecting the complexities of its design, even a term for the kind of craft that Sänger proposed eluded him. No vocabulary of space exploration existed when he first articulated the spaceplane concept, and commentators referred to the Silbervogel as an "antipodal" bomber, given its unusual flight path from one side of the Earth to the other. Others merely referred to it by the name of its designers ("Sänger-Bredt"); in the US, the term "skip-glide" later became associated with the Silbervogel's unique mode of flight, and the term "boost-glide," more generally, for any glider boosted above the atmosphere by a rocket. Building a functional boost-glide vehicle in the 1940s, though, was literally easier said than done. Had the German Luftwaffe commissioned the Silbervogel from Sänger and Bredt in 1941, Nazi Germany would have constructed the first vehicle to carry a human being into space—a human being who would have died on the very first test flight when the frictional heating produced by the Silbervogel's contact with the atmosphere melted its metal skin and incinerated the cockpit. Computational errors led Sänger and Bredt to underestimate the heating that the Silbervogel would endure during its flight; had they not done so, the pilot would have likely crashed the craft anyway.

Sänger alluded briefly, throughout his earlier published research and the 1941 report, to the difficulties that pilots would encounter attempting to fly Silbervogel, given its unusual flight profile and shape.[5] Indeed, a 1933 book outlining some of Sänger's ideas insisted that purely human piloting might well be impossible, as "the aircraft pilot will hardly be able to maintain a complicated mathematically defined flight trajectory in the few minutes that the ascent phase lasts."[6] Landing, meanwhile, might prove so hazardous that Sänger anticipated, in his 1941 proposal, that the Silbervogel could not be automated using the technology of the time and required "capable pilots" to facilitate it.[7] Decades later, American researchers examining vehicles like the Silbervogel determined that aircraft deriving lift from their fuselages instead of traditional wings ("lifting bodies") landed fast and were prone to weird yawing and other dangerous flight behaviors.[8]

Not surprisingly, the concept of a rocket-powered boost-glide vehicle—an airplane that could routinely fly into space powered by a rocket and glide back to a runway landing—remained unfulfilled throughout the war. Its construction required solving too many vexing technical problems; aerodynamic surfaces, engines, thermal protection systems, and navigation and control devices existed only as vague notions in 1941, and even by the end of the century, no aircraft with the Silbervogel's specifications flew.

What is perhaps more surprising is the remarkable durability of this impractical concept in the years that followed. While a variety of people and institutions in the US advocated for human space travel in the postwar period, efforts to develop boost-glide aircraft for military use were among the best funded—and ultimately the first to reach the hardware stage. The forces organized to support the development of such a vehicle did not see this craft as a tool for the peaceful exploration of space or interplanetary colonization, but as a hedge against rapid changes in military technology after World War II. This view is unsurprising, given the craft's origins as a "wonder weapon" whose value lay principally in its capacity for surprise attack. Of no discernible utility now, but of potentially critical value later, spaceplanes constituted an elaborate mechanism to ameliorate risk—a telos that required neither a working vehicle nor a mission to send it on.

Suicide Clubs and Space Travel

Of the seemingly futuristic technologies that emerged from World War II, the rocket-powered airplane was neither the most immediately practical nor the most dramatic.[9] Radar, nuclear weapons, penicillin, the digital electronic computer, the turbojet engine-powered airplane, infrared-guided air-to-air missiles, acoustically homing torpedoes, remotely piloted glide bombs guided by television cameras—developed in the US, Nazi Germany, or Britain during the war but made available to the US afterward—soon became the essential building blocks of postwar American defense technology. Other exotic wartime technologies like rocket planes demonstrated basic functionality but seemed many years away from practicality, even if their arrival seemed increasingly inevitable.

Postwar enthusiasm for rocket planes was not based upon wartime battlefield successes, but rather on extrapolation from advances in aviation that made rocket planes appear to be part of an inevitable future. From the wood-and-canvas gliders powered by gasoline-fueled motorcycle engines that Samuel Langley, Orville and Wilbur Wright, Glenn Curtiss, and other innovators experimented with at the turn of the twentieth century, the airplane had grown exponentially in size, weight, range, and complexity by 1945. Much of this transformation had occurred just before or during World War II: nations that operated wooden biplanes at the beginning of the war were flying jet fighters and high-altitude bombers with intercontinental reach by its end.[10] During World War II, airplanes carried supplies, photographed enemy territory, and dropped explosives capable of destroying the foremost technology of World War I—the battleship. Aluminum-bodied airplanes the size of school buses flew by the thousands in mass bombing raids over German and Japanese cities, at such high altitudes that crew members wore oxygen masks or sealed themselves in pressured cockpits that were little short of spaceships. Perhaps more than any other technology, the airplane revealed itself during the war to be not only a decisive weapon, but one capable of more destruction than any previously invented: mated with the atomic bomb, the airplane was a city-killer with a 3,000-mile range and the speed of a bullet.

The ubiquity and success of the airplane during World War II suggested to observers that it would play a central role in conflicts to come, and gains in speed and altitude made during the last war would be dwarfed by those of the next. Among the most promising new aviation technologies of the war was the turbojet engine, first described by the British engineer Frank Whittle in 1928, and employed by several nations during the war. Propeller-driven aircraft harness the power of piston engines to move an aircraft forward. The turbojet engine consists of a single cylinder with two sets of propeller blades (turbines) mounted inside it at both ends. The first turbine compresses the ambient air and drives it into the engine, where it mixes with vaporized fuel. Ignited, the fuel-air mixture creates a continuous blast of thrust sufficient to drive a second turbine (which powers the first) and produces enough thrust to propel the aircraft forward at great speed.

Another technology of the interwar period that appeared likely to hasten those gains was the rocket engine. Falling into a class of machines termed "reaction motors," all rockets generate propulsive force by ejecting some form of mass rearward: an action-reaction phenomenon described in Isaac Newton's Third Law of Motion. The reaction mass may be anything, but the heavier and faster it moves, the faster the rocket will accelerate; the hot exhaust gases produced by the rapid combustion of an organic fuel and oxygen are not particularly heavy, but they can be fast, if confined in a combustion chamber and released in a single direction. Black powder was the earliest rocket propellant: the greasy powder was a combination of three chemicals: a fuel (charcoal), an oxidizer (potassium nitrate), and an additive, sulfur, that lowers the temperature of ignition. As they carry their own oxygen with them in solid form, black-powder rockets work equally well in the atmosphere and in a vacuum and may be constructed to any size. The use of black powder in rockets emerged in China at least 1,000 years ago and likely preceded the related development of firearms.

Unlike the airplane, which was only forty years old at the start of World War II, solid fuel rockets utilizing gunpowder propellants had existed in global military arsenals for hundreds of years by that time, their value principally being their ability to deliver a significant mass of explosive a great distance without the use of heavy cannon. The earliest rockets were expendable wood or metal tubes packed with powder and strapped to a large stick that aided in the launching of the rocket and stabilized its flight. Rockets a few feet in length were common bombardment tools of the British military throughout the 1800s, with their employment in the siege of Fort McHenry during the War of 1812 immortalized in Francis Scott Key's poetry and the US national anthem.[11]

To launch a rocket above the atmosphere was not, even in the early twentieth century, far-fetched, but these rockets would, under the influence of gravity, decelerate and eventually fall back down to Earth. The dream of space enthusiasts was not only reaching space but traveling fast enough that the rocket never returned: it would either achieve a speed sufficient to overcome Earth's gravitational field and sail into interplanetary space, or it would coast endlessly around the planet, orbiting as an artificial satellite. Achieving the speeds necessary to launch a payload of significant mass into orbit around the Earth, or to escape it entirely, calculations first conducted by Konstantin Tsiolkovsky in his 1903 paper "Исследование мировых пространств реактивными приборами" (Exploration of Outer Space by Means of Rocket Devices) demonstrated that black powder was insufficiently powerful, and even the most chemically energetic fuel and oxidizer combinations like supercooled liquid hydrogen and liquid oxygen were barely strong enough. The amount of liquid fuel required to launch a vehicle 100 miles into the sky and accelerate it to the 17,500 miles per hour required for it to sustain an orbit around Earth required a craft so lean that propellant would constitute over 90 percent of its launch weight (high-performance fighter planes, by contrast, generally have a fuel weight of 30 percent to 40 percent, and airliners nearly 60 percent).

Tsiolkovsky and later rocket enthusiasts in several countries fixed upon the multistage rocket as a solution to the mass problem.[12] Ascending rockets would quickly deplete their fuel tanks, dragging the increasingly empty containers with them as they rose. Instead, one could assemble large rockets from stacked, smaller rockets—as each rocket stage depleted its fuel, explosives would separate it from the rest of the vehicle and the lighter, remaining stages would accelerate more quickly. Recognition of the possibilities of multistage rocketry helped establish a viable research program for the embryonic and cash-starved rocket research communities before and during World War II: instead of producing mammoth, liquid-fueled rockets able to reach the Moon, researchers needed only to produce smaller vehicles that might later be stacked to produce a lunar craft.

Although theorized earlier, rockets burning liquid propellants did not exist until Robert Goddard, a physics professor at Clark College (now Clark University) in Worcester, Massachusetts, built and flew small vehicles in the late 1920s and 1930s with grants from the Smithsonian Institution and the philanthropists Daniel, Florence, and Harry Guggenheim. Measuring ten to fifteen feet tall and consisting of metal tubing used to hold the tanks of liquid oxygen and gasoline and the vehicles' simple cylindrical engines, they were ingenious but not particularly impressive pieces of machinery.[13] In the US and throughout the world, independent groups of rocket engineers, sometimes in contact with their peers elsewhere, began to tackle the foundational problems of large rocketry, first by building viable motors (the essential element in any successful rocket) and gradually progressing to complete, flightworthy rocket stages.

The interwar period in the US and Europe was a kind of Golden Age of amateur rocket experimentation. Theoretical publications on the subject were increasingly commonplace, legal restrictions on the construction and possession of rockets—individually or by governments—were nonexistent, and a diverse and international network of experimentalists worked steadily to perfect rocketry's basic technology. Despite the paltry promise performance of early vehicles, designs for larger and more elaborate liquid-fuel rocket vehicles began to proliferate in the 1930s in technical papers and popular books on rocketry by the Romanian-born German engineer Hermann Oberth (among others), popular magazines, radio, and film, including the cinematic stories of *Buck Rogers*.[14] Prior to World War II, any organizations of the 1930s, like Oberth's Verein für Raumschiffahrt (spaceflight society) and a fledgling group at the California Institute of Technology (Caltech), shared a distinctly egalitarian ethos and focused their energies on the peaceful exploration of space.

Caltech proved to be an ideal incubator for such research, offering an aeronautical engineering program under the Hungarian Jewish émigré Theodore von Karman that supported graduate student research and tolerated the presence of local spaceflight enthusiasts unaffiliated with the university. These included the self-taught explosives chemist John "Jack" Parsons and the mechanic Edward Forman, who in 1936 became two of the founding members of a rocket club led by a graduate student named Frank Malina, which later became NASA's Jet Propulsion Laboratory (JPL).[15] Funding sources for their efforts were few in number and rockets proved fiendishly expensive to build; the club's early experiments were funded with a pile of $1 and $5 bills wrapped in newspaper, provided by a member of the group believed to have engaged in petty crime.[16]

Military services eventually offered steadier funding to rocket enthusiasts and demanded, in turn, a kind of focus and discipline that accelerated development efforts. Rockets had been in military use for centuries and could serve as effective bombardment weapons; the tension between peaceful and military uses of rocket technology, rather than emerging during the Cold War, was present almost from the rocket's inception.[17] By 1941, many international rocket groups had either been subsumed by or had contracted with their nations' military services to provide rocket weapons. Restricted in its acquisition of convention artillery by the Treaty of Versailles that ended World War I, the German army offered engineers funds to develop rocket artillery as a replacement. Even in Caltech's socialist-leaning "Suicide Club" (as Malina originally called the rocket group), contract work for the US government brought badly needed funds to the laboratory. JPL personnel spent much of the war developing small rockets to accelerate conventional airplanes during take-off before spinning off as the Aerojet Corporation, later a major government contractor. While most rocket enthusiasts embraced these new sources of funding, not all of them did: Malina participated in military projects throughout the war but drew a line at bombardment rockets, which he regarded as inherently immoral.

JPL and Aerojet pioneered a number of the technologies critical to early rocketry (and even to the development of the space shuttle), including storable, rubberlike solid rocket fuels (of the kind used to boost the space shuttle into orbit), and hypergolic liquid fuel and oxidizer combinations that burn on contact, obviating the need for ignition hardware.[18] (These found use in the shuttle's Orbital Maneuvering System [OMS]).[19] While JPL alumni continued to make significant contributions through the 1950s, though, others lost interest in rocket work. Parsons's enthusiasm for occult rituals and feuds with the science fiction writer L. Ron Hubbard (who stole money from him and slept with Parsons's wife) often distracted him from his rocket research. Killed in an accidental explosion in 1952, Parsons did not live long enough to see the applications of his rocket work.[20] Malina eventually refused military work, ran afoul of government authorities, and left both his rocket research and his US citizenship for life to become a civilian spaceflight advocate and artist in Europe.[21]

The hope of many rocket enthusiasts of the 1930s was not merely to build rocket artillery or a faster terrestrial aircraft. Instead, they sought to build a vehicle that could travel where contemporary airplanes could not: above the atmosphere that enveloped Earth, a border that von Kármán himself defined as 100 kilometers (62 miles, or 330,000 feet). At this altitude, wings could not generate lift and the air-breathing internal combustion engines that powered all existing airplanes suffocated; rockets of various kinds offered the only potential propulsive mechanism for airplanes operating at such heights. Not every rocket enthusiast favored a winged vehicle (American designers thought that German wartime rockets featured aerodynamic surfaces that were too large and heavy for the benefits they provided),[22] but others embraced the idea, for which no single individual can claim credit.

During the 1920s, occasional daredevils (most notably Germany's Fritz von Opel) strapped gunpowder rockets to crude gliders; this means of propulsion produced flame-and-smoke-filled demonstrations but offered little in the way of a practical solution to the problem of high-speed, high-altitude flight. Black powder was not particularly energetic as a rocket propellant, and poorly packed rocket motors often exploded when lit. Even when functioning properly, the rockets expended their fuel in seconds and could not be throttled or extinguished once ignited—fine for a bombardment weapon but not for a piloted aircraft, which might undertake hours of powered flight at varying speeds. The use of liquid propellants, whose combustion could be controlled by valves, offered one obvious solution, but this technology was embryonic at the time.

When Sänger articulated his concept for a liquid propellant boost-glide vehicle in his doctoral dissertation at the University of Vienna in 1928, the faculty refused to accept his "research" and demanded that he pursue a more conventional topic.[23] Despite this inauspicious beginning, the Silbervogel traveled unusually widely among the informal network of like-minded space enthusiasts. Sänger borrowed money to pay for the publication of his unsuccessful dissertation in 1933 (Oberth's press accepted the manuscript but refused to advance the printing costs),[24] but the resulting book, *Rakentenfleugtechnik* (rocket flight engineering) attracted little notice at first.[25] Shortly thereafter, though, a Caltech doctoral student named William Bollay hosted a seminar on the book in Pasadena that was attended by Parsons, who was captivated by the presentation. Other aeronautical engineers soon took notice of it as well.[26]

To many rocket engineers of the 1930s and 1940s, the rocket-powered airplane held obvious technical appeal over a rocket that ascended straight into the air: to rise vertically, a rocket needed to produce thrust significantly greater than its weight at liftoff and remain stable during its brief period of powered flight, when it would be laden with fuel as well as explosives. A rocket-powered airplane, however, required significantly less power to fly the same distance and was easier to control. The airplane's rocket motor would only need to push it forward, and the airplane's wings and aerodynamic control surfaces would generate the lift and steering forces required to keep it flying.[27] One immediate concern of rocket plane designers was that the mass fraction required for such a vehicle might preclude the use of wings altogether,[28] but rocket staging solved that problem. Sänger placed the first stage of the notional Silbervogel on the ground in the form of a rocket sled that would accelerate the craft to flying speed.

Like other first-generation rocket experimenters, Sänger devoted his time principally to engine design, seemingly the most complex technological hurdle.[29] Although Bredt later asserted that Sänger was a "most peaceloving [sic] person" who had deep misgivings about military work,[30] Sänger enthusiastically sought a research position in a German government laboratory to build and test rocket motors, and then, once funding for civilian rocket research dried up in 1942, turned his Silbervogel concept into a bombing aircraft to attract military funding. When this idea failed as well, he redoubled his efforts. Sänger's tireless revision of the concept to meet the approval of Nazi military authorities earns little mention in Bredt's 1973 account of Sänger's career; instead, Bredt's discussion of wartime work on the Silbervogel extends to only a single sentence, encompassing the hardships of 1944–1945, with this period of "total war" described chiefly as a tragic impediment to Sänger's spaceflight research.[31] Sänger and Bredt (who later married) were only two of many Axis rocket scientists who eagerly embraced military work while claiming not to have done so; indeed, a mixture of retrospective self-pity and revisionist pacifism was a feature of later Nazi rocket scientists, including those who fared better in the regime.

By lowering their expectations about winged flight, fellow German rocket researchers (and former Verein für Raumschiffahrt members) Oberth and the mechanical engineer Wernher von Braun produced liquid-fueled rockets that were more likely to work using available technology, but they still failed to produce a viable weapon, chaffed at military control, and sought to excuse their culpability for the crimes committed with their knowledge or under their direction. In 1944, von Braun's army-funded experiments at Peenemünde on the German coast yielded a missile (a rocket containing a guidance system) called the Aggregat (aggregate) 4 (A4), able to reach space and lob a one-ton explosive warhead several hundred miles from its launching point. Although standing forty-five feet tall and impressive when launched, the weapon was too inaccurate to strike military targets and very expensive to produce, soaking up funding and resources badly needed for more effective weapons (including most of Germany's available potatoes, needed to produce the alcohol that fueled it). Like the Silbervogel, the A4's relatively small bomb load and low accuracy eventually turned the missile into a blunt terror weapon to be used only for the murder of civilian populations, and it eventually joined the German arsenal as the Vergeltungswaffe Zwei (retribution weapon 2) or V-2.[32] The V-2 was, in fact, a war crime twice over, as its manufacturing facilities exploited tens of thousands of concentration camp laborers who were purposefully starved and murdered, a fact well known to those managing the program, including von Braun.[33] There is no evidence that von Braun ever protested the conditions inside the underground Mittlewerk plant that produced the V-2, and when von Braun eventually tired of the missile program, it was not for moral reasons, but professional ones.

Although a Nazi Party member and a Schutzstaffel (SS) officer, von Braun's loyalty was principally to himself and his dreams of spaceflight. His penchant for narcissistic and politically tone-deaf complaints about his government's refusal to devote sufficient resources to his space travel schemes eventually got the better of him. After a fellow party-goer denounced von Braun's casual criticisms of the Nazi government in 1944, he spent two weeks in jail before being released at the request of his superiors.[34] For von Braun, rocketry was a race, not to win the war, but to achieve piloted spaceflight. While impressed with Sänger's intellect, von Braun had opposed Germany's commitment of resources to the Silbervogel. The historian Michael Neufeld (2007) has speculated that von Braun feared not only a diversion of resources from the A4, but Sänger's likely success, which would deprive von Braun of the pioneering role in spaceflight that he craved.[35]

Among von Braun's most ambitious ideas was a plan to stack a winged, longer-range variant of the A4 atop an even larger rocket to create a two-stage intercontinental missile (the A9/10) sufficient to strike the US, or an even larger piloted space vehicle able to launch human beings into orbit around the Earth.[36] The design studies, which von Braun and his team began in 1939 and continued intermittently, duplicated Sänger's work and created a professional rivalry between the men. A prototype winged V-2, the A4b, appeared late in the war and remained experimental by the time of its conclusion.[37] Full German government support for the Silbervogel might have enabled Sänger to surpass von Braun's efforts.

Compared to von Braun's plans, the Silbervogel was far more ambitious it employed the traditional flight mode of a piloted airplane to provide guidance and stability but utilized a liquid-fueled rocket engine and would fly at heights and speeds for which little data existed as to aircraft behavior. Although seemingly offering a quicker path to high-altitude flight, liquid-fueled rocket planes would be challenging to keep in the air, keep cool, and safely land, problems that Sänger had only just begun to consider at the University of Vienna.[38] Even rocket propulsion could barely provide the energy necessary to propel airplanes into space: by adding wings, designers might make the craft's initial climb easier, but wings were useful only during ascent and landing. Beyond 330,000 feet, wings generate no lifting force at all, and their mass and the drag produced by the thinned atmosphere complicated the problem of achieving the altitude and speed required for orbital flight. Building a single vehicle capable of navigating both in an atmosphere and the vacuum of space was a sizable engineering problem, and even the earliest work by theorists suggested that the airplane was not the proper model for future spacecraft, despite the enthusiasm for this concept by many of the era's pioneers.[39] Undaunted, though, Sänger, von Braun, and others pursued such a craft.

The war produced no aircraft comparable to Silbervogel or winged A9/A10 and only a handful of rocket-powered planes,[40] including the Messerschmitt engineer Alexander Lippisch's Me-163 Komet (comet), a rocket-powered glider used briefly by the German air force as a high-speed interceptor to attack incoming American bomber formations. Experience with the Me-163 was not encouraging: the plane's stability as it accelerated toward the speed of sound (755 miles per hour at sea level) was notably poor, and a test flight at maximum speed (702 miles per hour) nearly destroyed the aircraft when it encountered wing-shattering aerodynamic flutter. Even worse, the rocket's propellants were toxic and unstable, and the craft's limited fuel supply gave it only a few minutes of powered flight. Ground take-off gobbled up fuel that was needed for combat; like the rocket planes that von Braun and Sänger proposed, a winged rocket plane of any utility needed some kind of rocket to catapult it into the air before the plane drained its own fuel tanks in the struggle to reach altitude. In the 1960s, the question of how to boost a rocket plane into space ultimately consumed almost as much effort as the question of how to design such a craft so that it could survive its journey out of the atmosphere and back.

The single-stage Me-163 was the limit of Germany's wartime achievements in rocket plane design. Experiments with similar but less well developed craft, like the Silbervogel and the Becham Ba-349 Natter (grass snake) ended with Germany's capitulation. Like the Me-163, Natter was a winged rocket intended to ascend quickly and attack Allied bombers, in the Natter's case with salvo of unguided rockets embedded in the craft's nose. Unable to land after reaching its maximum altitude, the Natter would split in two and the pilot would descend to the ground suspended from a parachute. In practice, the vehicle was a death trap; the first test pilot who attempted to fly it (a young, enthusiastic volunteer) died within seconds, and the Natter never found operational use.

The events of the Natter's first flight, including the sacrifice of a young officer in a virtually suicidal rocket weapon, bear some resemblance to the fictional plot of Thomas Pynchon's 1974 historical novel about the V-2, *Gravity's Rainbow*.[41] Charismatic male personalities, the sexual exploitation of women and men, and romanticized death feature prominently in the work, which bears an uncanny similarity to certain aspects of the German rocket program generally and to Silbervogel in particular. Indeed, a description of Sänger penned by Bredt in 1973 describe a rocket engineer who could have walked off the pages of Pynchon's novel: a man with "his own shy charm and glowing persuasive power," who "had succeeded in inspiring all his team and staff with great enthusiasm for himself and his plans—from the head of the testing (a typical example of the Austrian high aristocracy) down to the youngest canteen-help, (a pretty North German peasant girl)."[42]

The Silbervogel avoided the ignominy of failure by never progressing far enough in development to kill anyone attempting to fly it. In 1944, Sänger and Bredt elucidated their final design for the Silbervogel and the German government grudgingly approved the project, albeit with less money than Sänger had hoped (especially with a new aircraft requiring the development of so many radical technologies).[43] Evidence suggests that Sänger began work on the airframe (or possibly a wooden mock-up) by the end of the war: one aircraft, photographed by the US army at a German airbase in Lofer, Austria, in 1945, bears an uncanny resemblance to the Silbervogel. The image, published in 1947 and captioned to denote only an unusual unidentified aircraft, does not reveal the whole vehicle, though, and the development of such a mock-up likely represented only the initial stages of construction, if indeed construction had begun.[44] The Silbervogel's propulsion system would have been the most complex part of the craft,[45] and there is no evidence that its development was ever completed. Given the V-2's torturous history, the completion of a winged spaceplane riddled with design flaws would have taken Germany decades, especially given the resource shortages it faced at the war's end.

Rocket Planes in the Cold War

Experience with exotic flying weapons during World War II whetted the appetite of postwar American defense planners for of all kinds of propulsion technologies, but the panoply of devices with which the various combatant powers experimented was so large that it was never entirely clear which kind of aerospace technology was most practical, or even which planners were talking about at any given moment. Discussion of these new propulsion technologies used the words "jet" and "rocket" interchangeably to describe any propulsion system that worked by expelling exhaust gases at high speed,[46] and these terms were used as verbs, adjectives, and nouns to describe engines, the aircraft that they propelled, and the weapons that they carried.[47] Popular references through the 1940s became hopelessly confused: in journalistic use, a "rocket plane" could be a futuristic spacecraft, a turbojet-powered fighter, or a propeller plane firing rocket weapons.[48] The situation was no better in popular fiction: Rocky Jones, the hero of the eponymous 1954 television serial *Rocky Jones, Space Ranger*, flew a winged V-2 rocket dubbed the "Orbit Jet XV-2."[49] Sometimes poorly informed commentators attempting to clarify the distinctions made things worse, confusing the physical principles on which the various propulsion technologies operated and scrambling the names of their inventors (like one 1944 *New York Times* article that incorrectly attributed the invention of the turbojet to Tsiolkovsky instead of Whittle).[50]

Experts joined in the confusion: turbojet aircraft had already proved their value during the war and postwar military service branches rapidly acquired them; any association with them made the more fanciful rocket propulsion research seem more reputable. The founders of JPL performed no research on jet engines but used that term to describe their rocket work because jet engine technology was more established than rocket propulsion at the time,[51] and the laboratory hoped for government contracts.[52] JPL's development of solid fuel motors for rocket-assisted take-off, for example, went by the acronym "JATO," for "Jet-Assisted Takeoff," more often than "RATO," which more accurately described it. Eschewing the word "rocket" as the proper term for a reaction motor combusting a stored form of propellant and oxidizer likely helped sow confusion for the next two decades.

Public distillations of late-1940s army and air force planning frequently described a "wish list" of propulsion technologies from air-breathing turbojet engines and ramjets to bipropellant, liquid-fueled missiles: alternatives to the rocket plane offering better economy and performance.[53] The piloted, propeller-powered aircraft was the only means of delivering nuclear weapons in 1945, but the vehicles' limited range and low speed required bases close to intended targets and mission times measured in hours. With the arrival of practical turbojet engine-powered airplanes in 1944, aircraft doubled their maximum speed and altitude and became a potentially faster and more survivable means of nuclear weapons delivery. Piloted jet-powered aircraft could travel long distances on relatively small amounts of propellant, delivering nuclear weapons over intercontinental distances in a matter of hours. A cousin of the turbojet, the ramjet, offered even greater speed and high-altitude performance.

Long-range, unpiloted rocket weapons like the V-2 were capable of even faster speeds than jet-powered airplanes, and thus they presented another solution to the problem of nuclear weapons delivery. Turbojets require atmospheric oxygen, which limits the altitudes and speeds at which they can operate; rocket engines carry their own oxidizers, which diminish their fuel efficiency but increase their performance. Launched on arcing trajectories guided by radar and radio (a brief powered flight followed by a long period of coasting), ballistic missiles powered by rocket engines could strike targets in minutes rather than hours and required no wings or pilots to do so. Because they carried their own oxygen, they operated at altitudes and speeds that suffocated jet engines; at the altitude that ballistic missiles could reach, the Earth's thinning atmosphere would render aerodynamic control surfaces unnecessary. Gravity alone would ensure a hit on the target.

While jet engines worked more and more reliably after World War II, ballistic missiles were still an unproven novelty.[54] Amid the rapid technological changes of the early 1950s, the US air force (which became an independent military service branch in 1947) attempted to hedge its bets by trying to develop several bomb delivery systems, often combining the features of different flight paradigms in the same craft. New technologies of radio, radar, and inertial navigation (employing accelerometers to enable a vehicle to track its own path), for example, offered a way to turn piloted bombers into unpiloted cruise missiles. (Indeed, several crude weapons of this type were used during World War II.) Although seemingly a promising alternative to ballistic missiles, though, jet-powered cruise missiles were difficult to guide accurately over large distances and some variants hybridized several unreliable technologies at once (liquid-fuel rockets, ramjets, and electronic guidance), increasing their likelihood of failure. One proposed iteration of this concept employed a small nuclear reactor to heat the air traveling through it, giving a cruise missile the fuel range to fly anywhere on Earth.[55]

Cruise missile programs provided opportunities for wartime rocket pioneers like William Bollay to continue their work and develop technologies critical to later spaceplane development. Bollay enjoyed a long career in winged missile design; later commentators described him as more inventive than von Braun, declaring him to be the unheralded "genius" most critical to the success of American space rocketry.[56] After a stint in academia (including a post at Harvard University), Bollay settled at North American Aviation

(NAA) (North American Rockwell after 1967), one of several wartime high-performance aircraft manufacturers branching out into rocket flight. Throughout the 1940s and 1950s, NAA received Army and Air Force funding for the SM-64 Navaho, a cruise missile originally developed by Bollay based on von Braun's winged A4.

As the Air Force increased the missile's range requirements in the early 1950s, NAA responded with enlargements and improvements, including the addition of ramjet engines and auxiliary liquid-fuel rocket boosters strapped beside the craft instead of beneath it, anticipating later shuttle designs.[57] Struggling to integrate a variety of propulsion and guidance technologies, Navaho lagged badly in testing by 1956 and soon found itself eclipsed by other weapon systems. The engines and guidance systems created for it, though, had a second life in future NASA space programs, while NAA, after a change in corporate parentage, assumed the position of prime contractor for the space shuttle.

As Navaho struggled, parallel efforts by the Air Force to create pure ballistic missiles also hit technical obstacles that slowed development and demanded the expenditure of more funds than were available. The mating of ballistic missile technology with atomic weapons appeared to require either a substantial increase in the size of the rocket or a dramatic shrinking of the size of the warhead: this weapon system appeared so difficult to construct that enthusiasm for its immediate development faded. Guidance issues that plagued cruise missiles were possibly even greater for purely ballistic weapons: without any control surfaces or terminal guidance to the target, ballistic missiles would lob nuclear warheads in a particular direction with the expectation that they would coast, unpowered, for a predictable distance. This form of crude control might be enough to take a warhead close enough to a large city but appeared insufficient to strike particular targets within it.

With existing piloted airplanes unable to provide the speed, range, and invulnerability demanded by nuclear warfare—and unpiloted cruise and ballistic missile programs beset by propulsion and guidance problems—some members of the aerospace community found, in Sänger's spaceplane, what they believed to be a solution to both the nuclear delivery problem and the future challenge of piloted spaceflight. Dating the first appearance of a spaceplane in American discourse is difficult: postwar newspapers and magazines often described plans for winged spacecraft employing the war's breakthrough technologies,[58] but these ideas appeared earlier on the covers of science fiction magazines of the 1930s like *Astounding* and *Amazing Stories*. Despite early efforts at secrecy, images of American spaceplane concepts proliferated after World War II, cultivated by enthusiasts like von Braun and shared by *Collier's* magazine, Walt Disney, and other media outlets. Engineers working for established American defense contractors offered a variety of winged vehicles for civilian space exploration, a use for which no actual market existed, but this provided the designers with a platform to promote their ideas to the public, press, and Congress. At Goodyear in 1954, for example, Darrel Romick pitched a winged, reusable, multistage rocket named METEOR (an acronym standing for "*M*anned *E*arth-Satellite *T*erminal Evolving from *E*arth *O*rbit ferry *R*ockets") offering inexpensive flights into Earth orbit.[59]

A pilot himself, von Braun had moved in the direction of winged rocket vehicles during the latter years of World War II, authoring plans for long-range missiles and spaceships augmented by lift-generating wings. After the war, Operation Paperclip brought several Axis rocket experts to the US from Germany (or temporary detention in Britain). Awaiting work with the Army's rocketry program after arriving in the US, von Braun prepared technical documents about interplanetary space exploration and an abysmal science fiction novel about Mars travel using such winged craft, the massive technical appendix of which eventually attracted the interest of *Collier's*. The magazine ran a series of articles on interplanetary travel beginning in 1952, lavishly illustrated by Chesley Bonestall and Fred Freeman and reflecting von Braun's enthusiasm for enormous, winged rockets.[60] Although presses remained uninterested in von Braun's fiction, his technical appendix found a publisher, appearing in print in 1953 as *The Mars Project*.[61]

Despite his lack of firsthand experience with winged rocket vehicles, von Braun made extensive studies of their use in both navigating Earth's atmosphere and landing on the surface of Mars.[62] To withstand the accelerations produced at launch, designers had generally assumed that spacecraft crew members would sit in a reclined position in the nose of any Mars-bound rocket. A rocket lacking wings would then descend tail-first to its landing point, with crew members required to pilot such a landing while lying on their backs. In an era before effective computer control of space vehicles, this kind of piloting was too challenging even for skilled crews, jeopardizing a potentially yearlong Mars mission with a crash during its final moments. Spaceplanes, though, offered a conventional cockpit, front-facing windows, and horizontal landings that humans could perform.

The military utility appeal of this craft was obvious to the Air Force for institutional as well as technical reasons. Spaceplanes could, in theory, deliver even the largest nuclear weapons accurately with rocketlike speed because a human pilot would guide them to their targets. Such vehicles would compete favorably with unpiloted cruise and ballistic missiles, which could lob nuclear warheads halfway around the world but lacked the ability to precisely target, retarget, or recall them in the event of accident or changing circumstances, or what strategists referred to as "negative control." Under a pilot's steady hand, though, the lift generated by the spaceplane's wings would enable it to dip into the atmosphere purposefully to change its direction or distance of its flight; wings would also offer the craft a multitude of potential targets and landing sites.[63] The same spaceplanes that delivered nuclear weapons, furthermore, could, with augmented rocket boosters, achieve orbit, where they could complete a variety of other missions, like photographic reconnaissance. None of these applications required a winged craft, but to the engineering culture of the Air Force (and later NASA), winged vehicles would accomplish these missions with greater flexibility and reliability.[64]

The reasons for the spaceplane's appeal, though, were not merely technical. At a time when the future of most American military service branches hinged upon their ability to deliver nuclear weapons, spaceplanes offered a global delivery platform that still required the services of the thousands of human pilots who formed the core of the Air Force's officer class. Arguments for winged craft, the historian Roger Launius writes, often hinged on subtle cultural biases like these. Trained first as aeronautical engineers, for example, early rocket engineers in government service and private industry tended to favor airplanelike solutions.[65] Achieving orbital flight through spaceplane technology suited the Air Force, and it pleased the service's leadership to imagine that in a few short years, spaceplanes flown by Air Force officers would be the first piloted craft to reach space.[66] Research on spaceplane construction during the 1950s thus not only helped reinforce a particular spaceflight paradigm, it also fixed in the minds of engineers and planners the role of military pilots in any future space program.

Cloistered by the Army and tasked principally with transferring his knowledge of the V-2 to American engineers, von Braun had less immediate impact on the course of spaceplane research than his contemporary, Sänger, whose acolytes began to lobby the Air Force to develop the technology. Throughout the Cold War, enthusiasts for winged space vehicles like Sänger and Bredt pitched winged craft as an inexorable development in space technology, whose arrival had been delayed only by the prejudices of shortsighted engineers.[67] Wrote Bredt in 1973, "Of course, it may appear rather astonishing to the unaffected observer of technical development within the past 50 years that manned spaceflight did not evolve gradually and consistently from aviation; that with the ascent into the cosmos, one renounced the benefits of an atmosphere delivering oxygen and lift as well as with the concept of recovering the hardware."[68]

Bredt asserted that, in addition to the prejudice of engineers, the only reasons that "aerodynamic vehicles" did not displace ballistic missiles was their "high development cost" and the fact that they did not yet work. Instead of developing such craft in the 1950s and 1960s, the US and Soviet Union proceeded with ballistic missiles, a fact that Bredt almost absurdly attributed to the supposed success of Germany's wartime rocketry program, which was notable principally for its strategic failures.[69]

After the war, the US and Soviet Union gathered what technical information they could on the Silbervogel, gleaned through interviews of Sänger himself,[70] perusal of his writings (including at least one of the seventy printed copies of the 1944 design study),[71] and the testimony of captured German engineers briefed on Sänger's work during World War II. As knowledge of the craft circulated among the large but informal rocket and spaceflight community in the US, Operation Paperclip brought in several German experts with detailed knowledge of the program, including General Walter Dornberger, the head of Germany's wartime rocketry effort and an enthusiastic supporter of its National Socialist (Nazi) government.[72]

A German artillery officer during the Great War, Dornberger had received an engineering education during the Weimar period and exercised command over the Nazi program to develop long-range rocket artillery. After a brief consultancy with the US army, Dornberger and his young associate, the Peenemünde rocket engineer Krafft Ehricke, eventually found employment at Bell Aircraft Corporation in Buffalo, New York, where they began to expand and promote the piloted, boost-glide vehicle as a civilian exploration craft and military weapon system. Bell had ample experience with high-performance airplanes by the end of World War II, but its efforts were as extensive as they were uneven. Futuristic wartime fighter planes like the piston-engined P-39

Airacobra and the XP-59A Airacomet (the first American airplane powered by turbojet engines) had promised stellar performance but had not delivered in the field. The underpowered P-39 saw use in less-demanding environments and its successor, the P-63 Kingcobra, had a short production run; the XP-59A went unbuilt as a production aircraft due to its meager performance improvement over existing piston-engined aircraft. After World War II, Bell continued designing novel, high-performance aircraft for experimental and military use that were in many cases overambitious and trouble-prone, including helicopters, jet packs, and rocket planes.[73]

To support further work in this area at Bell, Dornberger spearheaded an effort to recruit Sänger in 1952 but was unsuccessful.[74] Offered the chance to join one of the few entities actually constructing rocket planes, Sänger instead choose to stay in Europe in a fruitless quest to interest French and German government agencies and aerospace corporations in the Silbervogel—which he now called the "space transporter"—but he was never able to attract sufficient interest to build the vehicle.[75] Sänger died in 1964, exhausted and, as Bredt recounted, emotionally broken; Bredt followed in 1983.[76]

In Sänger's absence at Bell, Dornberger cast himself as the "Father of the Spaceplane." In 1954, Dornberger published a remorseless memoir of his wartime rocketry activities, which disingenuously cast himself as both Germany's foremost rocket pioneer and the premier postwar advocate for piloted rocket planes.[77] Unrepentant at the circumstances of the V-2's manufacture and use, Dornberger recalled his role in the development in the A4b and a future spacecraft vehicle bearing an uncanny resemblance to the more ambitious Silbervogel:

> We were well on the way to solving a problem which, together with high-altitude research, was the first I had set myself to tackle after the war: the landing after a flight into airless space. We had taken a long stride forward in developing the first intermediate stage preceding the spaceship. Rocket aircraft could cover long distances in the upper stratosphere, at heights of 12 to 16 miles, at incredible speed, and still land safely. If only we could succeed in maintaining full rocket thrust just long enough to ensure that we reached this height at very high supersonic speed, flying horizontally and in the right direction, and then either go into a glide or switch on a low-thrust cruising motor using very little propellant, why, then we should be able to bridge thousands of miles in an economically feasible manner. Such were the ideas that occupied our minds in 1944. If we could realize them in practice we might hope to enrich international traffic, a few years after the war, with newer and bigger models. This revolutionary form of transport could never be rivaled for speed and range by normal propeller or jet aircraft.[78]

Perpetuating a popular postwar fiction among German rocketeers, Dornberger insisted that his goal throughout the conflict had been civilian spaceflight, which he hoped to pursue "after the war" (i.e., once Germany had conquered the Allied nations and exterminated all non-Aryans). Elsewhere in his memoir, though, he made it clear that space exploration had seldom entered his mind during the conflict: "Employment at the front seemed more urgent to us in wartime than this research, important and interesting as it might be scientifically. I was firmly resolved to use our rocket for it after the war, but for the moment the rapid march of events left me little time for dreams of the future."[79]

Slowly at first, starting in January 1946 and then moving more quickly, a series of experimental, mostly rocket-powered planes (X-planes), many built by Bell, began flight testing above the dry lakebed of California's Muroc Army Airfield (later Edwards Air Force Base). Bell's XS-1 (later X-1), with Major Charles "Chuck" Yeager at the controls, became the first aircraft to exceed the speed of sound in 1947, ushering not only a boom in supersonic aircraft design, but a renewal of enthusiasm for piloted, rocket-powered aircraft (figure 1.1). With early turbojet engines not yet capable of generating the thrust for supersonic flight, the X-1 employed a liquid-fueled rocket motor that was one of the last developed by Robert Goddard before his death in 1945.

The pilots who flew X-planes, from the Air Force, Navy, and the civilian National Advisory Committee for Aeronautics (NACA), NASA's principal predecessor agency, expected the performance of these craft to increase until one of the vehicles went into space.[80] Each year, X-planes further nudged speed and altitude records, leading to design studies for even more capable vehicles. While Bell was a leader in this technology, larger aviation firms like the Boeing Company and NAA—companies with better political connections and a better record of wartime success—dominated the postwar contracting environment and proved eager to offer anything Bell promised. One study, authored by Dornberger in 1954 for the NACA, eventually led to a contract (not with Bell, but with NAA) to produce the X-15, the first X-plane to climb to attitudes so devoid of atmosphere that the Air Force and Fédération Aéronautique Internationale considered their pilots to be spacefarers.

As Dornberger hoped, successful X-planes led to the inception of several programs to weaponize the technology in the form of military craft able to spy or bomb from space, as well as an increasing number of popular representations of civilian spaceplanes in science fiction, news magazines, and

Figure 1.1
Bell Aircraft Corporation's XS-1 in flight in 1946. NASA photo.

semischolarly publications produced by von Braun, the German émigré and space enthusiast Willy Ley, and others with passing knowledge of German wartime research.[81] Boosted into space by either a ballistic missile or large airplane, the sleek, delta-winged, piloted rocket bombers proposed by Bell, Boeing, and other contractors in the 1950s would streak across the upper atmosphere to any location on Earth within forty-five minutes—too high and fast to be intercepted—and bomb with a precision that pilotless cruise missiles like the Navaho and prototype ballistic missiles could not yet match.

Intrigued both by Sänger's creation and by Bell's postwar design studies, the Air Force solicited proposals for several piloted, high-altitude military rocket planes in the 1950s, with Bell offering some of the most ambitious designs. Bell, in fact, had not waited for a formal design competition to begin its work. In 1952, unprompted, it had submitted a design for a "Bomber Missile," dubbed "BoMi," consisting of a rocket powered bombing airplane carried into space by an even larger rocket carrier plane, with both reusable.[82] The ambitious design anticipated that of several future space shuttle configurations, but it was far too ambitious for its time. Three years later, an unfunded Air Force solicitation to contractors produced a more modest Bell proposal, the Rocket Bomber, or "RoBo," which placed a small, winged space glider atop an expendable liquid-fuel rocket, another configuration that anticipated later designs. The term itself was an anglicized version of the German acronym "Rabo" or "RaBo" (short for *"Raketen-bomber"*), which Sänger and Bredt had sometimes used to describe the Silbervogel.[83] Bell and other airplane makers received small amounts of Air Force funding for their design work, and Bell studies so far exceeded those of other manufacturers that the company, perhaps naively, assumed that it would receive the prime contract to build a spaceplane as soon as the Air Force authorized it.[84]

Meanwhile, Dornberger pitched unclassified variants of his own vehicle designs as civilian spaceships. One variant proposed the creation of a passenger version of his two-plane design: a large, winged carrier with a small spaceplane mounted on its back. At launch, both planes would sit atop a platform tail-down, both their engines igniting to send the combined stack skyward. The fuel tanks of the larger plane would supply its engines and, for a time, those of the smaller craft. Having expended its fuel before the smaller plane, the larger plane would separate from the smaller plane and the smaller one would continue its journey into space, crossing oceans in a fraction of the time that turbojet-powered aircraft could. Even more, both portions of the craft would be reusable.[85] Although largely ignored at the time, these and comparable designs became the basis for NASA's preferred configuration of the space shuttle a decade later.

In 1956, though, US government enthusiasm for funding a large passenger rocket was nonexistent. The administration of President Dwight D. Eisenhower had sought to reduce government expenditures by relying upon nuclear deterrence instead of conventional forces, under the doctrine of "Massive Retaliation" (nicknamed "The New Look" after Christian Dior's popular fashion trend of 1947). Massive Retaliation, ironically, spurred the creation of an expensive new generation of aircraft and missile weapons, but Eisenhower felt no comparable need to facilitate the development of civilian space rockets, seeing public expenditures on this research as violating his inconsistent commitment to thrift. The events of the following year, though, forced a reexamination of all American rocket programs (both military and civilian) and created an urgency for their acceleration that had not previously existed: on October 4, 1957, a Soviet ballistic missile blasted a probe the size of a beach ball into Earth orbit, altering the course of spaceplane development for decades.

Sputnik and Dyna-Soar

By early 1955, a competition was already underway among scientific agencies in the US to launch the first artificial Earth satellite into orbit. The vehicles to carry it included off-the-shelf military ballistic missiles and scientific "sounding" rockets used for atmospheric research, staged or clustered to produce a launch vehicle with enough thrust to launch a small, instrumented package into low-Earth orbit.[86] Launched atop a variant of the R.7 liquid-fuel missile designed by Sergei Korolev, Sputnik 1 was itself technically unremarkable, but its launch vehicle was a monster, large enough to hurl a hydrogen bomb over transoceanic distances. The launch pointed to the maturity of the Soviet Union's quest for an intercontinental ballistic missile (ICBM), but it also humiliated American space researchers who were on the verge of launching a satellite of their own. While orbiting a simple scientific satellite with an American ballistic missile became a critical short-term goal for the US, the Air Force continued to press for an orbital spaceplane, concerned about apparent Soviet gains in this area as well.

Earning somewhat less public notice than the front-page news of Sputnik's launch was a near-simultaneous claim in one Soviet news publication (reported in the *New York Times* on February 18, 1958) that the Soviet Union would soon complete development of a 10,000-mile-per-hour rocket plane capable of achieving orbital altitudes.[87] No such rocket plane existed, but like many instances of Soviet propaganda, the authors of the original piece had chosen to exaggerate a true story rather than fabricate an entirely new one. Soviet forces had found a copy of the 1944 Sänger-Bredt report at the conclusion of World War II, and translations circulated among Soviet designers in the months that followed.[88] The mathematician Mstislav Keldysh, who later assumed a prominent role in the Soviet space program as president of the Soviet Academy of Sciences,[89] assessed the Silbervogel in 1946 and began to design a Soviet version of the craft.

Reviewing Sänger's work, Keldysh, like many of his German and American counterparts, recoiled at the fuel consumption of the bipropellant liquid fuel rocket engines intended to power Silbervogel and began working instead on a vehicle powered by an air-breathing ramjet engine like the one powering

Navaho.[90] The US air force had undertaken similar work, hoping that air-breathing engines powered not by aviation kerosene, but by liquid hydrogen, might produce the propulsion efficiencies necessary to make the Sänger-Bredt spaceplane work. Throughout the late 1950s, projects to develop reusable, air-breathing, winged space launch vehicles continued under various names (including the "Aerospaceplane"), and while none produced flight hardware, several major US aircraft contractors produced impressive design studies.[91] Like Keldysh's phantom rocket plane, though, the Air Force designs of the 1950s required technologies that did not exist even fifty years later.

Months before the launch of Sputnik 1, the Air Force combined several underfunded rocket plane programs into a single multiphase weapons system, named "Dyna-Soar" (short for "Dynamic Soaring"). The Soviet launch of Sputnik 1 prompted a reassessment of American space programs, and the Air Force responded by soliciting renewed Dyna-Soar proposals from defense contractors in March 1958. By May, only two companies appeared to still be in the running: Boeing and Bell.[92] Bell produced the most sophisticated design but lacked Boeing's stellar record in producing successful military aircraft (and the goodwill it enjoyed within the Air Force); as occurred in previous and subsequent aerospace contracting, the Air Force solution was to rig the competition by prodding its preferred manufacturer to build a competitor's design. Boeing assented to the scheme, winning the contract to develop Bell's spaceplane for the Air Force.[93]

Boeing's revised entry into the Dyna-Soar competition (renamed the X-20 by the Air Force in 1962)[94] duplicated Dornberger's work. Intended to be lofted into space atop an Air Force intercontinental ballistic missile (ICBM), Dyna-Soar would, in an improvement on the Silbervogel, fly high and fast enough to reach orbit, returning to Earth only when its pilot fired a rocket motor to slow the vehicle just enough to cause it to scrape the atmosphere, further decelerating the craft for a gliding landing on a conventional runway. Critical to the craft's flight would be a metallic thermal protection system (TPS) sufficient to absorb and dissipate the 3,000 degrees of frictional heating on the craft's surface as it slammed into the Earth's atmosphere at over 17,500 miles per hour: a technical challenge that had gone unsolved in Silbervogel's original design. Although likely to fly long after the first artificial satellites launched by ballistic missiles, Dyna-Soar would combine the speed of a ballistic missile with the flight characteristics of piloted airplanes, including the ability to deliver nuclear weapons at great distances under precise human control. Early versions of Dyna-Soar would be research and reconnaissance vehicles, but in its final form, the craft would be a nuclear bomber with global reach.

Military analysts less vested in the future of the airplane were less impressed with the X-20—its capabilities appeared duplicative of those of other weapon systems, and debates about the utility and likelihood of success plagued the program for the next five years. While some subsystems for the craft were completed as early as 1959 (making the X-20 the first American piloted orbital spacecraft to reach the hardware stage), other critical components lagged and design problems quickly arose. Like most space vehicles, the X-20 experienced mass growth during its design process that called into question the proposed means of launching it. Originally intended to be flown on the Glenn L. Martin Company's SM-68A Titan I ICBM during early test launches, program managers in 1960 requested the use of the more powerful LGM-25C Titan II, which was still two years away from completion.[95] Subsystems, particularly the structures and mechanisms to protect the craft from reentry heating, remained unfinished, and virtually all of the X-20's novel technologies demanded a drawn-out testing program of suborbital launches and orbital test flights using enhanced booster rockets (eventually including the as-yet-unbuilt Titan IIIC) and new upper stages (figure 1.2). Even with constant annual appropriations, Air Force planners expected the first orbital flight as late as 1966,[96] and operational flights of the system no earlier than 1974.

Spaceplane in Search of a Mission

Critically, an actual reason for the X-20's development (other than to demonstrate the craft's own capabilities) never materialized in the years that followed the program's inception in 1958. Rather, existing weapons programs—the Titan II and the solid-fueled Minuteman I ICBM, the unpiloted Corona[97] spy satellite program—proved so successful that the X-20's unique applications began to shrink. The same concerns seemed to have undermined the early Soviet spaceplane projects as well, particularly plans for winged, reusable, multipurpose spaceships proposed by the Soviet designer (and Korolev rival) Vladimir Chelomey. Korolev, noted the historian Asif Siddiqi, dismissed the spaceplane concept as a "'circus,'" during one 1959 briefing by Chelomey's deputies. "'Why do we need such a system at the present time?' Korolev continued. 'Right now all this is

Figure 1.2
Air Force captain Edward J. Dwight, Jr., poses with a model of the X-20 and its Titan IIIC launch vehicle at Edwards Air Force Base, California, in 1963. The cancellation of Dyna-Soar prevented Dwight from flying into space on an Air Force vehicle. Although chosen for advanced aerospace training at the request of the administration of President John F. Kennedy, Dwight was not selected for NASA's astronaut program, a disappointment that Dwight attributed to NASA's racial climate.[98] US Air Force photograph, courtesy of the National Archives at College Park.

fantastic. In space right now it's necessary to solve [more specific] goals, for example . . . , photographing the far-side of the Moon.'"[99]

While the Air Force studied many uses for the craft, the X-20's capacity for runway landing alone ultimately proved most important to General Curtis LeMay and other Air Force leaders pushing for the spaceplane. Reviewing the program in 1963, President John F. Kennedy's secretary of defense, Robert McNamara, concluded that an enthusiasm for runway landing, rather than the Air Force's operational needs, actually drove the project. The only capability that mattered was the X-20's employment of Air Force pilots to fly it and Air Force runways to land it on: the destination and payload were irrelevant. Anticipating criticisms of the space shuttle, Clarence Geiger noted McNamara's concerns in his 1963 Air Force history of the program:

> After hearing presentations of the X-20, Gemini, and Titan II programs in the middle of March, Secretary McNamara reached several conclusions which seemed to reverse his previous position on the experimental nature of the Dyna-Soar program. He stated that the Air Force had been placing too much emphasis on controlled reentry when it did not have any real objectives for orbital flight. Rather, the sequence should be the missions which could be performed in orbit, the methods to accomplish them, and only then the most feasible approach to reentry.[100]

The response of the Air Force, provided by Harold Brown, its director of defense research and engineering, was paradoxical, arguing that an assessment of the X-20's capabilities (and hence, its need for runway landing) wouldn't be possible until the X-20 proved that it could land on runways. This did little to satisfy McNamara, and when later analysis revealed that a new Air Force spacecraft based upon NASA's Project Gemini capsule could accomplish most of the X-20's objectives more cheaply, the X-20 lost McNamara's support entirely. Even the craft's one additional advantage—the ease with which it could be made reusable—ultimately provided little added impetus to continue the program: reusability offered little benefit when the need for the craft's existence remained so weak and budgets for disposable craft remained ample. With the Air Force unable to demonstrate a viable need for Dyna-Soar (and headlines in national newspapers likening the spaceplane to an extinct reptile),[101] the Kennedy administration canceled the X-20 in 1963.[102]

The lessons of the abortive Dyna-Soar program were ultimately almost completely forgotten in the years that followed, to the detriment of the eventual space shuttle program. During the late 1950s and early 1960s, the Air Force and its contractors struggled to find a powerful, reliable booster for the X-20 and a TPS offering physical robustness in addition to heat resistance. Amid these technical problems, planners also wrestled with the fundamental illogic of the system, particularly the degree to which function and purpose were subordinated to the ultimately meaningless requirement of runway landing. Even reusability, second only to runway landing as the holy grail of future spacecraft design, offered little actual projected benefit: funding for national security projects was generous, expendable launch vehicles were already the norm, and few believed that the efficiencies in operating a fully reusable spaceplane were worth the cost of building it.[103]

Conclusion

As a case study in technological choice, the story of the spaceplane's ascendance in the 1950s and demise in the early 1960s is a remarkable example of the limits of wishful thinking in the design of space vehicles. Aerospace engineers believed that the culmination of their profession would be a craft that melded the rocket and the airplane, and their conviction that such a vehicle would emerge was sufficient justification to push for its development despite its technical challenges and lack of utility. For Air Force, NASA, and contractor engineers working to put men into space in airplanes, the spaceflight achievements of the 1960s were a constant reminder of the superiority of ballistic missiles and capsules, as well as unpiloted vehicles of all kinds. While the X-20 stumbled over launch vehicle and thermal protection issues, the Air Force's and NASA's expendable rockets surpassed those of the Soviet Union in the delivery of satellites to low Earth orbit and beyond. Advances in communications, navigation, and computing removed the need for an onboard human presence in orbit, and as the accuracy of nuclear missile guidance improved due to advances in microelectronics and computing, piloted bombing aircraft of all kinds were gradually deemphasized in in favor of a triad of delivery systems supplementing piloted aircraft with unpiloted land- and submarine-based ballistic missiles.

Neither were other nations or private companies interested in investing in spaceplane technology: the commercial spaceflight market in the 1960s was embryonic and the US government enjoyed a monopoly on space launch capability in the Western world. NASA occasionally lent low-cost satellite launchers to allied nations to develop their space industries but exporting a technology to make space access affordable—using technology that might be repurposed to launch nuclear weapons—was not a priority of US foreign policy. Nor could American spaceplane designers look to Soviet accomplishments as a reason for the US government to fund its own: stymied by technical problems, Soviet researchers allowed their spaceplane program to languish. In 1965, the Soviet Union revived its program and then, ten years later, accelerated it in response to the American space shuttle program, but by then, Soviet research was twenty years out of date and produced little of value. Analogous to the X-20 in design and function, the Mikoyan-Gurevich MiG-105 experienced similar difficulties and was canceled before flying in space.[104]

Thwarted in his efforts to build the Silbervogel in the US, Dornberger's career was ultimately a pale imitation of that of his subordinate, von Braun, who occupied high-profile US army and NASA positions through the Space Race of the 1950s and 1960s. Both worked to build military rocket planes while promoting their use for civilian space exploration, but one worked in a struggling private firm while the other found himself at the center of a growing government agency. A more culpable figure due to his acknowledged wartime control of the V-2 program and utter lack of remorse, Dornberger enjoyed some productive years at Bell, but he never achieved von Braun's success or celebrity.

Ehricke, a propulsion expert, spent little time under Dornberger at Bell, moving first to Convair in 1952 and then to other aerospace companies to pursue various spaceflight projects. Although best known professionally for his work with engines powered by liquid hydrogen and oxygen (later critical in the development of the space shuttle), Ehricke achieved public notice only for his theory of the "Extraterrestrial Imperative." Drawing on Malthusian fears of population growth and modern concerns over environmental degradation, Ehricke suggested that interplanetary space travel would be necessary to sustain future life on Earth.[105] Working in North American Rockwell's Space Division in 1971, at the height of the debate about NASA's space shuttle, Ehricke preached in speeches and articles about the potential role of space mining and space manufacturing to solve the world's resource needs, activities that would be facilitated by an efficient spaceplane.[106]

If the advocates of the X-20 learned anything during this abortive program, it was that runway landing alone was not sufficient to justify the development of a boost-glide space vehicle. Reusability, and the imagined cost savings that it provided, offered an additional inducement, but with budgets for space exploration climbing in the mid-1960s, cost cutting remained of minimal interest to either the Air Force or NASA. Instead, spaceplane advocates would have to think bigger, even at the risk of making fools of themselves. When the first orbital spaceplane was actually built, it was not a robust bombing aircraft like the Silbervogel or a small space fighter plane like the X-20 or MiG-105; rather, it was a complex kludge whose purpose had questionable merit and whose vulnerability to physical and political forces was obvious.

2 Star Clipper and Other Fantasies, 1963–1969

Of all of the "wonder weapons" explored by German engineers during World War II, the Silbervogel rocket plane perhaps best epitomizes the idea that it is often the weapons that show the least chance of actually working that attract the greatest popular interest.[1] Impossible to build using the technology of World War II, the boost-glide bomber transfixed a certain group of postwar aerospace enthusiasts in several countries, all of whom periodically pressed civilian and military leaders to revive the concept. These efforts led eventually to the space shuttle by the National Aeronautics and Space Administration (NASA): not a copy of the Silbervogel—an impressive, if half-conceived, weapon of war for a fascist dictatorship—but a nervous democracy's supposed instrument of science and diplomacy. A direct descendant of the Silbervogel but not a simulacrum, the shuttle was a creative reinterpretation of the spaceplane concept, developed after several previous iterations of the boost-glide concept had failed to demonstrate utility or practicality, even amid the flush budgets of the Cold War.

Congress's establishment of NASA in 1958 and the agency's growing responsibility for America's embryonic human spaceflight program altered the course of existing rocket plane development projects. Prior to that point, the rocket pioneer and later NASA Marshall Space Flight Center (MSFC) director Wernher von Braun had imagined a permanent human presence in space and visits to nearby planets. In a gigantic, doughnut-shaped, Earth-orbiting space platform, dozens of spacefarers would perform scientific research and prepare for deep-space adventures. Like a cosmic train station, this platform would dispatch pilots, astronomers, and engineers on voyages to the Moon or Mars, fuel and provision their vehicles, and communicate with them as they ventured into the solar system.[2] A far cry from the isolated polar explorers of the nineteenth century, the inhabitants of von Braun's orbiting wheel would come and go daily via Earth transfer ships, the design of which shifted between rockets, rocket planes, and some hybridized combination, as von Braun had proposed as early as 1952.[3] Von Braun's writings and those of comparable futurists often used the term "ferry" to describe such craft; the term "space shuttle" emerged in popular use around 1959, generally in lowercase, as a generic term for any spacecraft that achieved routine flight from Earth to space.[4]

The Space Race with the Soviet Union, however, ruined von Braun's plans for elaborate space transportation infrastructures, shifting attention away from complex vehicles like space shuttles, permanent orbiting space stations, and interplanetary flight, and toward Earth-orbital spaceflight employing ballistic capsules and existing military missiles. Although many engineers at NASA supported spaceplane research, the agency's political mandate to challenge Soviet space supremacy demanded an acceleration of programs that were more likely to match Soviet achievements in the short term. The Soviet launch of Sputnik 1, though initially accelerating boost-glide programs, eventually pushed the slow-moving rocket plane projects into the background in favor of capsule-based space vehicles offering fewer technical challenges and a faster timeline for completion.

Even after the demise of Boeing's X-20, a variety of contractors and government institutions throughout the 1960s pitched plans for spaceplanes, hoping to revive the concept utilizing a grab bag of supposed benefits—efficiency, flexibility, and military utility, among others—mixed in various proportions to suit the political climate. Competition in space with the Soviet Union hurt rather than helped these efforts: the need for quick successes in space exploration mitigated against space vehicles requiring novel technologies or extensive research and development programs. Once it became clear that the US would not need a spaceplane to accomplish a piloted landing on the Moon (as in the Air Force's proposed Lunex program),[5] the urgency to immediately develop such a craft faded. Undaunted, spaceplane advocates of the 1960s instead argued that to move beyond the achievements of Project Apollo, the US would need to make flying in space as cheap and effortless as airline travel, a goal that they argued could be met only by a reusable vehicle that had some ability to glide in the atmosphere. Faith, not reason, guided this advocacy: no spaceplane had ever worked, and the only drawback to capsules was that they were so simple that they bored the engineers building them. The need to craft a justification for building a spaceplane forced its advocates to propose the vehicle as the only practical solution to a range of disparate problems, each of which barely existed, and all of which begged for different answers.

Space Pilots for Spaceplanes

Efforts by engineers to design spaceplanes in the 1950s and early 1960s moved in tandem with similar efforts to train the people to fly them, particularly in the Air Force, whose aviators were already involved in the testing of high-performance aircraft at Edwards Air Force Base. These men expected to be the first class of space travelers once rocket-powered X-planes managed to escape Earth's atmosphere, and despite various fanciful popular representations in the 1940s and 1950s and NASA's brief consideration of nonpilot crews,[6] little evidence exists to suggest that US government agencies and military service branches ever seriously expected anyone but a trained aviator to fly America's first generation of space vehicles. As early as the mid-1950s, research conducted by the Air Force's aeromedical community, including director of research at the U.S Air Force Air Research and Development Command, Brigadier General Don Flickinger, virtually assumed that military aviators were optimal candidates for space crews, and the question assessed by aviation physicians and others was more likely to be which individual pilots would qualify, rather than whether pilots were appropriate at all.[7] Not waiting for the completion of spacecraft, the Air Force set to work creating the procedures for selecting future space crews and soon established a roster of test pilots to serve in this role, many drawn from the X-plane program. Selection of the men likely to serve as space pilots (the word "astronaut" was not yet in use) preceded the fabrication of the spacecraft that they would fly, so program managers built vehicles around crew skills rather than selecting crews to suit the preferred hardware.

North American Aviation (NAA) unveiled the X-20's immediate predecessor, the X-15, in October 1958, and the black, arrowlike craft first flew the following year amid speculation that it would eventually serve as a reusable orbital spaceplane.[8] In 1963 (the year that the X-20 was canceled), the X-15 succeeded at becoming the first piloted, winged craft to reach space, though only briefly. Dropped from a Boeing NB-52 Stratofortress carrier aircraft over Edwards Air Force Base, the rocket plane's liquid-fueled rocket engine boosted the craft and its pilot, Joseph Walker, to an altitude of sixty-six miles, but with insufficient speed to achieve orbit. After reaching its maximum altitude, Walker glided the craft by hand through the atmosphere to land at Edwards.[9] The technological demonstration provided evidence of the workability of the boost-glide concept, but plans to send the X-15 into orbit atop rocket boosters created for Navaho never left the design stage.[10] (A more ambitious proposal to launch the X-15 into space atop an early Saturn launch vehicle also amounted to little.)[11] With simpler approaches to space travel already yielding results, additional investments in spaceplane technology appeared unnecessary.[12]

With Dyna-Soar not expected to reach operational flight for several years and the Soviet Union moving rapidly toward human spaceflight in 1958, Air Force attention turned to launching a human being into space quickly, in a metal capsule the size of a telephone booth mounted atop an existing Air Force ICBM.[13] This new, cheaper, and more expedient spacecraft dispensed with wings, large windows, and other accouterments of airplanes that added needlessly to the weight of the craft. Instead, it consisted of only a pressurized chamber in the shape of a sphere or cone, mounted atop a missile and guided into space and back by radio signals generated by ground-based computers. These would execute engine firings for launch and reentry of the spacecraft remotely, with the occupant merely providing biomedical data throughout the flight. Instead of gliding to a piloted landing on a runway, the capsule would deploy parachutes and land in the ocean, where its buoyancy would keep it on the surface until plucked out of the water by a helicopter. Without the need to fly horizontally through the air, the capsule could dispense with wings. Wings were useless in the vacuum of space, anyway, and magnified the thermal issues associated with a spacecraft's recovery: thin, sharp wings suited to supersonic flight through the atmosphere were difficult to protect from overheating during reentry, and, capsules, arguably, were safer during virtually all phases of flight.[14]

Existing military pilots were the only candidates the Air Force considered to crew its new space capsule despite the vehicle's lack of resemblance to an airplane: in June 1958, the Air Force selected nine respected test pilots as potential crew members for its "Man-In-Space-Soonest" project. When NASA assumed control of human spaceflight in August 1958, the Air Force canceled the MISS project, but it was influential in the selection of astronauts for Project Mercury, NASA's civilian space program that duplicated the Air Force's capsule scheme.[15] At first, NASA balked at limiting its spacefarers to pilots—capsules did not require any piloting at all, and a piloting requirement would eliminate tens of thousands of worthy physicians, scientists, and engineers who might make good use of the space environment. Shortly after the agency's founding in 1958, NASA's Space Task Group even undertook a brief civil service recruitment effort that, had it not been canceled weeks later, might have led to the hiring of a more diverse group of explorers and scientific specialists as the first American spacefarers.[16]

Daunted by the prospect of an open, public search for the first American space travelers, though, NASA quickly changed course, and at the insistence of President Eisenhower, it drew its first astronauts (loosely defined as "star voyagers") from the all-white, all-male ranks of its military test pilots, just as the Air Force had anticipated (figure 2.1). Members of NASA's Space Task Group had originally considered the phrase "space pilot" for the new professional group, ultimately modifying the word "aeronaut" to create what they

Figure 2.1
The "Original Seven" NASA astronauts of Project Mercury pose with a model of their launch vehicle and spacecraft (the truncated cone at its tip). Early astronauts (all active-duty military test pilots) were uncomfortable with the vehicle's configuration and the capsule's splashdown landing mode. Front row (left to right): Virgil "Gus" Grissom, Scott Carpenter, Donald "Deke" Slayton, and Gordon "Gordo" Cooper; back row (left to right): Alan Shepard, Walter "Wally" Schirra, and John Glenn. NASA photo.

thought was a neologism.[17] Even after NASA introduced the word "astronaut" into popular parlance, though, astronauts and journalists continued to use the more specific term "space pilot," with its connotations of a defined professional skill set.[18] NASA did not need actual pilots to fly its Mercury capsules, but it wanted the reliability and discretion that military test pilots could provide; the would-be astronauts wanted the opportunity to be the first Americans in space, but not at the expense of their professional identity as aviators.[19] Within NASA's astronaut corps, a military piloting culture quickly took root and bled into the public sphere through press reports about the men. Under the astronauts' influence, NASA's space "capsules" became "spacecraft," crewed by "pilots."[20]

World War II, having ended only fifteen years earlier, provided not only the pilots who made space exploration possible (including NASA astronauts John Glenn and Deke Slayton, as well as the X-15 pilot Robert White),[21] but a vocabulary with which to describe men willing to endure such risks. NASA astronauts emerged from a military piloting culture and flew jet planes constantly to maintain their proficiency. The American public seized upon astronauts as combat professionals, risking their lives in a grand international contest.[22] Depictions in print fiction, film, and television emphasized astronauts' clean-cut image and stoic acceptance of danger, common to popular depictions of military aviators.[23] Their lives were replete with risk and sacrifice, but also alcohol, philandering, and foolhardy forms of recreation: astronauts often played chicken with each other, stretching the fuel reserves of their jets to see who could fly farthest without crashing.[24]

Emerging from the ranks of military pilots, NASA's astronauts were often the most enthusiastic supporters of spaceplane technology. Some had experience with rocket plane programs or had worked closely with those who had, and they found the mode of flight of rocket planes safer and more satisfying. Particularly galling was the splashdown phase of a capsule's flight, with the crew members awaiting rescue and risking drowning "like a bag full of cats" until helicopters and ships arrived to pluck them out of the water.[25] At first, though, the astronauts' concerns were ignored. The Space Race of the 1960s was a political problem, not an aeronautical one, and the astronauts that the US needed were not pilots, but professional managers and public relations consultants. When the Air Force recruited astronauts for its Dyna-Soar program in 1962, it designated the men as "pilot-engineer consultants,"[26] an inelegant but more accurate description of their role. Astronauts gradually realized what NASA's actual needs were and sought to establish their management reputations both with the agency's engineers and more established test pilots, who looked at Project Mercury with a mixture of contempt and alarm. They also labored to make space capsules more like the airplanes they knew best, with greater options for manual piloting, even at a time of increasing automation in flight vehicle controls.[27]

NASA's human space record in particular during the 1960s quickly validated the decision to table spaceplanes: within a few months, the capsules of Project Mercury matched Soviet space achievements, culminating in Gordon Cooper's daylong Mercury flight in 1963. The Project Gemini spacecraft, an enlargement of the Mercury vehicle, surpassed Soviet capabilities by the end of 1965 and demonstrated workable spacewalking and rendezvous techniques by the end of 1966. Using winged technology for Project Apollo's voyages to the Moon became unthinkable: with mass allowances tight for Project Apollo's lunar voyages, a structural component that would be useful only upon returning to Earth made little sense. Apollo's use of rendezvous and docking techniques would enable the crew to conduct the final portion of their journey to the Moon's surface in a lunar module optimized for flight in a vacuum, with the crew members returning to Earth in Apollo's conical command module capsule. These plans worked remarkably well, and while parachute recovery and water landing provided NASA with a few moments of concern for crew safety, this approach to recovering space vehicles proved relatively straightforward, given the US navy's large surface fleet and global operations.

NASA simply had no need for a winged space vehicle throughout the 1960s, although engineers and astronauts often tried desperately to create one. One noteworthy example was the effort, undertaken by Francis Rogallo and NASA's Langley Research Center (and enthusiastically supported by some NASA astronauts), to equip space capsules with an inflatable wing to facilitate runway landing. Rogallo had hoped to employ the system for Project Apollo but faced resistance from Apollo engineers; a more sympathetic response from Project Mercury engineers led to plans to use the concept on Gemini. Stowed during launch and orbital flight, the triangular wing would deploy during landing, and the spacecraft would hang from it. Too small to accommodate landing gear, the modified Gemini would carry folding landing skids—equipment that, when deployed after reentry, enabled the capsules to glide a runway landing instead of dangling by a parachute over the Atlantic Ocean.[28] By utilizing a wing that deployed only once the capsule had safely descended into the

Figure 2.2
NASA astronaut Gus Grissom (left) and test pilot Milton Thompson (right) pose with the Project Gemini paraglider test vehicle (Paresev 1-A) on Rogers Dry Lakebed at Edwards Air Force Base, California, in October 1962.

atmosphere, the parafoil offered NASA the opportunity to create a true hybrid craft, with the operation advantages of both capsule and wing designs but fewer of the disadvantages of either. On paper, the system offered significant efficiencies over the X-20, in that its lighter wings would be present only when they were needed.

Development of the Gemini paraglider proceeded from 1961 through 1964, enjoying the support of several NASA astronauts and test pilots who tested full-scale mock-ups of the vehicle in various stages of wing deployment and landing cables (figure 2.2). While the wing might have become practical with sufficient investments in time and money, Gemini had less of the former than the latter. Although conceptually simple, the use of the parafoil required the solution of two distinct problems: the deployment of the inflatable wing as the craft descended, and the handling of the new vehicle on landing, each of which produced testing mishaps that destroyed vehicle mock-ups.[29] Slippage in deadlines and repeated testing failures pushed the use of the paraglider into Gemini's later missions, and then led NASA to drop it entirely.[30]

The spaceplane's seeming last gasp in human spaceflight of the 1960s thus ended without the flight of a winged vehicle, though Rogallo's patent later established the basis for the sport of hang gliding.[31] NASA's enthusiasm for spaceplanes had seemingly run its course, but even as work on space capsules progressed, some astronauts remained hopeful. The X-20 and the Gemini paraglider had attracted fans like astronauts Michael Collins and Gus Grissom, and new astronauts continued to join NASA from the flight test community, occasionally transferring from rocket plane programs. The success of capsules did not appear to materially change their orientation toward lifting into flight with their wings: astronauts' continued advocacy in the 1960s was another critical factor in the spaceplane's eventual embrace by NASA in 1968.

Scramjets and Shuttles

The rapid development of missile and space capsule technology in the US during the early 1960s made elaborate plans for spaceplanes unnecessary: the US, by 1963, possessed proven space launch vehicles and capsules offering robust capabilities with few technological unknowns. Yet even before the US had won the Moon Race, many engineers within NASA and private industry favored a resumption of the more elaborate "von Braun paradigm":[32] the launching of a large, Earth-orbiting space station and a variety of planetary missions, supported by a spaceplane. Sensing that the Air Force's interest in spaceplanes had waned, advocates of boost-glide vehicles abandoned military justifications for them and pressed their case instead that a spaceplane would hasten the construction of a larger civilian space infrastructure. When support for massive civilian space infrastructures proved scarce, spaceplane advocates then argued that a reusable spaceplane would lower the cost of all space operations and create new commercial, exploration, and military opportunities in space. These efforts attached the feature of operational efficiency to lifting flight, utilizing the former as a selling point for the latter.

Although deemphasized in favor of capsules, spaceplane concepts had never entirely disappeared after Dyna-Soar's cancellation, particularly in experimental flight, military, and NASA circles. Astronauts continued attempting to add lifting surfaces to their capsules in Project Gemini, and contractor studies for space shuttles (most likely lifting vehicles of some kind) that could carry 100 passengers into orbit by 1975 were under review at NASA in the early weeks of 1963.[33] Minutes from a meeting of government agency representatives at MSFC on January 24 confirmed the participants' conviction that while the "mission requirements which would indicate a clear-cut decision to develop highly reusable space launch vehicles do not yet exist," such vehicles "will be brought into service eventually," especially in connection with the support of a large, orbiting space station.[34] Despite Dyna-Soar's cancellation, the Air Force's program to develop a recoverable orbital launch system continued through 1964.[35] This enthusiasm for a failed technology is striking—by 1963, spaceplanes had become a self-fulfilling prophecy, thought necessary largely because of their perceived inevitability.

By this time, MSFC under von Braun was rapidly developing an expendable heavy-lift launch vehicle—the Saturn V—capable of lobbing an entire space station or translunar craft into orbit intact on a single flight. While the three-stage, liquid-fueled rocket was larger than necessary for most Earth-orbit missions, a smaller, two-stage Saturn IB vehicle shared an upper stage with the Saturn V, and both were modular vehicles that could fly in two-, three-, or even four-stage configurations as the mission demanded, creating rockets with cost-effective component redundancies across the Saturn system. The economics of such a vehicle over a one-size-fits-all space shuttle, though, failed to persuade engineers already captivated by the spaceplane concept. NASA and various contractors instead seized on the sizable outlays for Saturn as an excuse to propose an entirely new space architecture built around a smaller, more reusable spaceplane that would be cheaper to operate. "Are these monsters becoming too heavy and their launch facilities too gigantic—and too expensive?" Walter Dornberger inquired in a 1965 magazine article criticizing the Saturn system and advocating for spaceplane research.[36] It was a question that nobody was asking (Saturn vehicles were modular, and series production was actually lowering their per-unit cost), intended to be resolved by a solution that Dornberger had been pushing since 1944.

Attaching reusable spaceplane construction to a space station and interplanetary flight was counterintuitive: a large booster would orbit a space station or Mars vessel with the fewest possible launches. Spaceplane proponents, though, argued the opposite: each of these grand space ventures required a larger number of launches carrying smaller payloads, each of which was too inconsequential to justify the expense of a Saturn launch. Among the most absurd explications of this idea was Dornberger's insistence that "every man in a space station circling the globe this side of the Van Allen Belts should have a fresh egg on his breakfast table every day brought up from Earth."[37] The logic of Dornberger's argument was strained to the point of ridiculousness, assuming a need for frequent launches of reusable supply vessels to provide orbiting astronauts with small luxuries that most Earthlings did not enjoy.

Whatever the justification, NASA and many of its contractors agreed with Dornberger that the next American human spacecraft should be a reusable, Earth-orbiting vehicle with a cargo bay capable of launching people and the components of larger spacecraft into orbit. While such a system need not be an airplanelike craft, most designers assumed that it would be. Expendable ballistic missiles shed so many disposable and unrecoverable components—engines, tankage, guidance systems—that barely a fraction of the launch vehicle's dry mass even returned from orbit. An airplanelike orbiter paired with an airplanelike, reusable booster, though, could eliminate this waste and obviate the need for a naval fleet to recover returned capsules. By combining lifting flight and reusability with a large cargo capacity, such a spaceplane would, in theory, also lower the cost of delivering cargo to Earth orbit. Upon the shuttle's return to Earth, a spaceplane could even carry back malfunctioning satellites for repair and redeployment on subsequent missions, perhaps making the spacecraft's operations profitable. These applications, though, were all entirely notional, deriving from fantasies of spaceflight's future rather than a clear-eyed assessment of its current needs.

In response to solicitations by NASA future project offices, defense contractors that had competed for Dyna-Soar contracts (including Boeing, the Convair division of General Dynamics, the Lockheed Corporation, and the Martin Marietta Corporation, the last formed by the merger of Martin and American-Marietta Corporation in 1961) proposed reusable spaceplanes throughout the mid-1960s. Although hazy about the specifics, the NASA requests generally envisioned a vehicle capable of flying into low Earth orbit 100 times carrying ten tons *or* ten passengers—much less than the payload of the shuttle NASA eventually commissioned.[38] Critically, NASA's invitations for proposals excluded a specific call for a lifting vehicle, but they did emphasize reusability as the principal system goal. In practice, most contractors recognized that NASA wanted a lifting craft and was unlikely to support a reusable capsule-based option. Still, some contractors (like Chrysler Corporation) offered impressive reusable capsule designs. The automotive giant's Space Division had worked successfully on the Saturn IB, but its shuttle designs seldom attracted significant interest despite meeting NASA's cost and function objectives.[39]

The Air Force and NASA continued to study reusable spaceplanes even as expendable rocket boosters and space capsules racked up impressive achievements in space through 1966, convinced that once the Moon Race had abated, the design hiccup created by the ballistic missile launch paradigm would end as well. Dornberger's and Bell's preferred craft (and that of Republic Aviation[40] and numerous other contractors dating back to 1960) had been a winged, two-stage, fully reusable, piloted spaceplane, piggybacked atop a piloted booster employing an exotic new air-breathing engine technology to achieve hypersonic speeds within the upper reaches of the atmosphere: the supersonic combustion ramjet (i.e., the "scramjet").

The scheme was entirely notional and exceeded even NASA's ambitious shuttle plans; again, Bell found itself pushing too hard for an innovative solution to a problem that barely existed. Air Force colonel Donald Heaton, assigned to NASA headquarters under the assistant director of propulsion within the Office of Space Flight Development in 1960, was one of many experts skeptical of the fanciful new spacecraft. Periodically solicited by contractors on NASA's interest in hypersonic vehicles, Heaton recalled demurring as a matter of principle. The technology, he declared, was all but useless, even if anyone managed to build it:

> Providing the Air Force or somebody provides the billion or more dollars that would be required to bring this vehicle into being and if it proves to be a good vehicle we would be very happy to use it, but . . . our studies have left us quite bearish on the likelihood that an air breathing winged recoverable launch vehicle would be used sufficiently or frequently to warrant the highly expensive development program which would be involved and highly speculative chance that such a launch vehicle would be a completed success in the foreseeable future.[41]

In 1965, the idea of a scramjet-powered spaceplane was too revolutionary for NASA or the Air Force to contemplate: scramjet propulsion remained unproven through the end of the twentieth century despite a series of experimental programs intended to validate it, and the earliest successful test of a scramjet engine did not occur until 2001 (see chapter 6).[42] Dornberger, though, refused to back down, and like his former colleague von Braun, he began to assail the work of other engineers in an effort to advance his own projects. Unwisely, in a July 1965 address at the University of Tennessee, Dornberger dismissed any spaceplane solution short of the scramjet approach, including the type of space shuttle that NASA eventually commissioned: a more modest craft powered by rocket engines and launching and landing from installations in Florida and California. Dornberger declared:

> Before I leave the shuttle spacecraft, let me just mention some studies which have been conducted by industry on the recoverable, reusable space transporter. Looking at the development costs, the rocket-powered shuttle is certainly cheaper and development will be shorter, because the rocket engines for the vehicle have already been developed. However, its takeoff weight will increase almost 200% when it has to carry on board from the very beginning all of the oxygen required, and the dimensions of the craft will limit its operational use to a few well-prepared takeoff and landing sites. And this is *not* what we want.[43]

Only the air-breathing spaceplane would offer the economies of scale necessary to sustain its use, Dornberger insisted, and while he may have been correct, NASA had no intention of developing such a craft. (Influential NASA associate administrator George Mueller, for example, ignored proposals for air-breathing spaceplanes entirely.) Thwarted, Dornberger retired from Bell in 1965 and played no subsequent role in the development of the space shuttle; instead, he returned to West Germany, where he died in 1980.

Although a reusable spaceflight infrastructure would save money if optimistic projections proved accurate, the space shuttle in 1965 was an idea rather than a specific vehicle architecture. Each proposed approach required the solution of different technical challenges and offered different cost structures. Designs with low operating costs had high development costs, and vice versa. The solutions most likely to succeed at reducing overall costs were not always the most innovative. Technical analyses tended to favor incremental improvements to existing hardware over radical new developments and unproven technologies like lifting bodies, hypersonic engines, and unusual staging systems. Meanwhile, a small but persistent segment of the design community lobbied for a new generation of inexpensive but reusable ballistic missiles that would return to Earth using parachutes and be recovered from the ocean. Many of these proposals were grouped informally under the category of Big Dumb Boosters: rather than a step forward in design, they were a self-conscious step backward, offering reduced performance at a greatly reduced cost, with some measure of reusability. Some of these were pressure-fed liquid fuel rockets,[44] like SpaceX's later Falcon series (itself based upon NASA research of the 1990s), which would dispense with the elaborate pumps that forced propellant into engines in favor of tanks of pressurized gas. Others imagined a new generation of massive, solid-fueled rockets with virtually no moving parts; such rockets were cheap and reusable but had never been used in piloted space vehicles due to the inherent risks that they presented.

Efficiency analyses of shuttle options assumed significant increases in the number of space launches in the near future; otherwise, shuttlelike space vehicles made little economic sense. NASA managers were often unpersuaded by cost estimates and promoted infrastructures based on their own peculiar design preferences. Mueller, the legendary spacecraft designer Maxime Faget, and other shuttle proponents saw lifting boosters and crew vehicles as NASA's future: a means of allowing the spacecraft to land on conventional runways, speed turnaround time between missions, and dispense with the armada used to retrieve previous space vehicles and boosters from the ocean. Lifting craft would be more expensive than capsules but would likely be more comfortable for professional and nonprofessional astronauts to fly in reducing training time and costs and opening the space program to a more diverse pool of participants, including paying customers from industry (and possibly even tourists). Other technical additions, like jet engines (to enable shuttles to cruise through the atmosphere after returning to Earth) would add cost but provide versatility, safety, and comfort.

By 1967, NASA and the Air Force had narrowed their respective searches to some kind of reusable craft powered by conventional rocket engines and capable of runway landing—neither NASA nor the Air Force were certain what it would look like, but initial assumptions seemed to rule out Bell's air-breather, a reusable capsule, a new Big Dumb Booster, or a reusable glider placed atop an expendable rocket. Other measures short of full reusability, including the use of an expendable fuel tank on an otherwise reusable vehicle, remained under consideration. NASA's preferred configuration for a spaceplane—a fully reusable, winged, liquid-fueled shuttle with a piloted, liquid-fueled fly-back booster—presented the most expensive option for initial design and construction, but it also promised lower life-cycle costs if it operated as expected.[45]

Engineering analyses suggested that in the absence of specific objectives for the shuttle, a less expensive space vehicle would fulfill all projected launch needs, even those required to support a space station or interplanetary exploration. Advocates for the shuttle's economic benefits thus based their lobbying upon a series of technological capabilities that the craft would supposedly offer over existing vehicles—especially runway landing—which would create new space traffic. The Air Force had attempted the same argument to justify the X-20's development, but runway landing was no more necessary in the case of the space shuttle than it had been with Dyna-Soar.

Arguing without evidence that demand for launch services was elastic, NASA seized upon reusability as the secret to lowering launch costs, and upon runway landing as the secret to reusability. Although lifting flight, reusability, and cost efficiency were independent features (NASA and the Air Force had researched reusable capsules, expendable winged vehicles, and inexpensive space launch systems using solid-fueled rockets launched from airplanes), NASA's decision to conflate the three features produced an "all-or-nothing" mentality that ultimately led to a very expensive preferred configuration. With initial estimates of the development costs for the program exceeding $12 billion, the shuttle would cost half as much as the entire Moon program but would realize significant savings over time once launch traffic increased.[46] The winged vehicle envisioned by NASA would be capable of multiple flights with only minor refurbishment between missions; the reduced capital expenditures on each flight would create large economies of scale once flights became frequent.

Left unstated in most spaceplane proposals at the time was the source of the projected growth in space launches. Expendable launch vehicles were comfortably managing the existing load of weather, communications, and spy satellites upon which the nation relied, and an infrastructure to send more people and cargo into space appeared to be a solution in search of a problem. To NASA, though, the space shuttle's capabilities could not be assessed in isolation, only as a component of what Mueller described in 1969 as a "Space Transportation System": a constellation of space vehicles that would usher in a new era in space exploitation and exploration. These vehicles would include a space shuttle, a "tug" to carry personnel and equipment between low and high-Earth orbit, a space station, a nuclear-powered interplanetary craft, and a variety of robotic vehicles.[47] Of these, the shuttle was the most critical, as it would enable the launch of the other components. Buttressed by the needs of space station construction and interplanetary voyages, the shuttle's missions would soon explode in number.

Although never endorsing the idea wholeheartedly, NASA even promoted the space shuttle as a passenger craft or tourism vehicle. One NASA rendering, titled "Space Shuttle Missions: Global Transportation," depicted a shuttle carrying dozens of civilian passengers.[48] Others imagined voyages to a Hilton-owned space hotel.[49] (One iteration of the concept would lodge 100 guests in rooms buried beneath the lunar surface to protect them from meteors and radiation.)[50] Concept art from one noteworthy 1966 shuttle proposal, the Star Clipper from Lockheed, similarly depicted the craft in space with its large payload bay packed with coach airline seats.[51]

The Star Clipper, in particular, offers a prime example of the surrealism that surrounded NASA's call for reusable vehicle studies. By the mid-1960s, few defense contractors had the experience with space vehicles possessed by Lockheed, a venerable Southern California airplane-maker with a growing business in high-performance aircraft, rocket stages, and spy satellites.[52] Throughout the 1950s and early 1960s, though, Lockheed had steered clear of experimental rocket planes, with the famed designer Clarence "Kelly" Johnson skeptical that such craft would teach engineers much of value in the design of practical airplanes.[53] Lockheed's team also recognized that the passenger plane analogy was inapposite as far as the space shuttle was concerned: no large craft could fly into space and back without shedding at least some mass, and any effort to make such a system entirely reusable would cost more than NASA was willing to spend. Determined nonetheless to offer a viable entrant in NASA competitions, Lockheed designer Maxwell Hunter advocated for a partially reusable glider with an airplanelike fuselage and a jettisonable, expendable fuel tank, a design feature that ultimately found its way into NASA's final space shuttle design, although he received no credit for the innovation. Indeed, the design appeared to be one of the stronger contenders at first, which made its disappearance in the years that followed all the more noteworthy (see chapter 3).[54]

Given the enthusiasm that some American defense contractors showed toward space shuttle design studies, one might assume that the companies' leaders saw the potential for tremendous profit. For the contractors, though, space shuttle studies were paper projects lacking significant funding commitment, and companies seldom aggressively promoted their work publicly despite the fact that Air Force and NASA space shuttle competitions were open and publicly reported in the popular press.[55] Lockheed, a proven space manufacturer and the largest US defense contractor during the 1960s, made only a cursory mention in corporate annual reports of its efforts to design a post-Apollo spacecraft, never promoting these efforts significantly.[56] (Lockheed institutional histories leave the Star Clipper out entirely.)[57] Lockheed's rocket and space business instead revolved around contracts for successful, high-cost military programs like the Polaris submarine-launched ballistic missile and the Corona series of photographic reconnaissance satellites. These were active programs with a continual need for new hardware, paid for by a blank check from the national security establishment. Until 1968, space shuttle design studies were aspirational projects with no congressional mandate.

For the contractors entering NASA and Air Force design competitions, shuttle plans were more a hedge against future downturns in defense business than a path toward riches. Civilian human spaceflight projects were subject to substantial congressional and executive branch oversight, risks were high, and profit margins were slim. Nobody would get rich off the space shuttle (unless they were rich already), and defense contractors never invested significant amounts of their own money developing designs or demonstrating the feasibility of hardware. "Try before you buy" may have been the norm in military aviation contracts of the time, but the sums required for a space shuttle were too great, and the rewards too slim, for anything more than token corporate investments until production contracts were at stake. These were unlikely so long as Project Apollo progressed toward its goal of making a lunar landing by 1970, but NASA, as enthusiastic as ever about spaceplanes, aggressively promoted the idea that contracts to build spaceplanes would be forthcoming.

While final approval for the shuttle did not come for many years, a 1967 symposium at NASA headquarters conducted by Mueller (who remained a key proponent of the spaceplane concept throughout the debates that followed)[58] assessed a variety of proposed vehicles as follow-on craft for Apollo, leaning heavily toward lifting craft. Another symposium followed in January 1968;[59] that year, the Air Force issued contracts for design studies under its Integral Launch and Re-entry Vehicle (ILRV) program; NASA's Manned Spacecraft Center (MSC) in Houston, Texas, and MSFC in Huntsville, Alabama, assumed control over the program by the end of the year, redesignating the current round of design studies as "Phase A" of a multiphase program that NASA renamed the "Space Transportation System (STS)."[60] The ILRV/STS Phase A studies marked the official beginning of what became the dominant human spaceflight infrastructure for the US for the next forty years: they were eventually followed by more detailed configuration proposals (Phase B), detailed designs (Phase C), and finally, a production contract (Phase D) for a fleet of vehicles.[61]

Although known officially by the "STS" moniker throughout its lifetime, studies and proposals both inside and outside NASA during this period increasingly described the craft as a "space shuttle." The use of the term in a speech by Mueller before the British Interplanetary Society in August 1968 at the time of the Phase A competition probably sealed its acceptance among contractors competing for the production contract:[62] "The Space Shuttle is another step toward our destiny, another hand-hold on our future. We will go where we choose—on our earth—throughout our solar system and through our galaxy—eventually to live on other worlds of our universe. Man will never be satisfied with less than that."[63] Having received official endorsement, the term "space shuttle" became forever associated with the IRLV proposals and subsequent plans for a reusable, piloted launch vehicle.

Throughout December and early January 1970, the Office of Management and Budget (OMB) feuded with NASA acting administrator George Low over NASA's insistence that the post-Apollo launch vehicle always be described using the "space shuttle" moniker, as NASA believed that naming the embryonic project would prevent the new administration of President Richard Nixon from killing it.[64] Within a year, though, Low's replacement, James Fletcher, was having second thoughts about the name: while politically expedient, "space shuttle" was too utilitarian.

A memorandum by Fletcher to the White House on December 30, 1971, contemplated whether the space shuttle program would eventually receive a name comparable to Project Apollo or its predecessors Gemini and Mercury, but the shortlist prepared by Public Affairs, shuttle development staff, Low, and Fletcher himself were uninspired. The goofy-sounding "Skyclipper" and "Skyship" topped the list, a deviation from NASA's tradition of naming space projects after characters in classical mythology. ("Pegasus" and "Hermes" rounded out the top four.[65]) NASA made little progress on the naming question during the shuttle's development: on August 12, 1976, Fletcher wrote to NASA assistant administrator for external affairs Herbert Rowe to ask what name the agency had decided on for the shuttle program, then rapidly approaching the rollout of the first flight test vehicle.[66] In a reply six days later, Rowe wrote that he could not recall the result of the previous deliberations, but he believed that, having rejected the alternatives, NASA had chosen "space shuttle program" as the project's official moniker, with individual orbiters receiving "patriotic names" like "Constitution, Independence, America, etc."[67] Naming the shuttle would prove almost as vexing as figuring out exactly what it would look like.

Saturn Rockets and Squashed Lightbulbs

Aerospace firms bidding for contracts in Phase A in 1968 offered a bewildering array of single-stage and multistage rockets, capsules, and gliding craft, ranging from the traditional to the ridiculous.[68] With each successive phase of design studies, flight architectures sometimes changed dramatically. Emphasized through the process, sometimes to the detriment of other capabilities, was the need to build a space vehicle that would handle well enough to be flown by human pilots, a requirement that always meant airplanelike wings, cockpits, and control surfaces. Otherwise, neither NASA's multiple field centers nor its contractors had achieved even a basic consensus on the size, shape, or performance of the new vehicle.

Many shuttle proposals bore the indelible marks of their idiosyncratic authors, especially when NASA laboratories joined the design competition, fearful of being left out of the program. Under Faget's direction, NASA's Langley Research Center continued to press for the purest iteration of the shuttle: a stacked pair of winged rocket planes with straight, simple wings that would make up for in reusability and flying ease what they lacked in payload capacity and other features. Faget worked with X-15 and Project Apollo contractor North American Rockwell (with NAA and Rockwell having merged in 1967) on such a design: a craft dubbed the "DC-3" (in honor of the pioneering passenger plane) (figure 2.3). Aware of NASA's obvious preference for this configuration, other contractors followed suit with similar designs.

Fearing that NASA might leave MSFC out of shuttle development, von Braun advocated for the new, reusable, winged booster, but only if MSFC secured the contract to design it. Barring that, von Braun pushed for the reuse of MSFC's Saturn V as a space shuttle booster, ensuring a continued role for MSFC in the shuttle program even if NASA opted against the DC-3. The latter option was attractive for a variety of technical reasons: whether used as a booster for a new glider or a reusable capsule, von Braun argued, the expendable Saturn launch vehicle was an extant, proven technology with ample room for growth. The Air Force had reflown a Gemini capsule in an earlier test flight and Apollo command modules could be similarly reused with the replacement of their ablative heat shields. The per-mission cost savings of a reusable capsule atop an expendable Saturn launch vehicle would be less than that of a fully reusable shuttle, but a production line for new Saturn rocket stages bought in quantity would likely lower their cost in time.[69] Engineers from the MSC in Houston prepared similar plans, and some contractors responded in kind, offering Saturn-derived shuttle concepts in an effort to flatter the biases of NASA's in-house design teams.[70]

MSFC and Boeing were so confident in the reusability of existing Saturn hardware that they even explored the possibility of recovering the ostensibly expendable Saturn rocket stages and flying them again: even when dunked in seawater, the smaller Saturn IB's first-stage H-1 engines fired in static tests. As early as 1964, NASA studies indicated that with the addition of parachutes and some navigation equipment, the entire first stage of a Saturn IB might be recovered, refueled, and reused, and there was little reason to believe that larger Saturn V first-stage (S-IC) vehicles couldn't be reused as well, with minor modifications.[71] Studies indicated that retrofitting a S-IC with parachutes and other recovery systems had a minimal effect on the launch vehicle's payload capacity and would likely save the agency "hundreds of millions of dollars" in fabrication costs, but such plans received relatively little attention despite the tremendous advantages they offered.[72] NASA headquarters never allowed alternatives to a new winged shuttle to advance; indeed, NASA's refusal to

Figure 2.3
The gold standard of reusability for the space shuttle throughout the many rounds of contractor proposals was the "double-fly-back" booster and winged orbiter. Launched vertically and flown by human pilots, both the large booster and the orbital vehicle would land on runways: the booster section immediately after launch and the orbiter at the conclusion of the mission. NASA's Langley Research Center collaborated with the North American Rockwell Space Division of Downey, California, to produce several designs, including the one depicted here. NASA photo, courtesy of NARA, National Archives Identifier: 17409729.

take Saturn use seriously for the shuttle despite congressional pressure to lower costs, had, by 1964, already frustrated Edward Z. Gray, NASA's director of advanced piloted missions. "We find it embarrassing to the agency as a whole," Gray noted in one memorandum, "to implement a study program wherein study contractors expend resources to respond to our request and we then do not follow through."[73]

Among the likely candidates for a reusable spacecraft that would launch atop a reusable Saturn rocket were gliding vehicles without wings: bulbous, round-bottomed lifting bodies that generate aerodynamic forces sufficient to glide without conventional wing structures.[74] Eliminating wings from gliding designs was appealing for several reasons: in every phase of spaceflight except landing, wings were a hindrance: their weight, drag, and structural vulnerabilities greatly complicated spacecraft design. Well before the announcement of the ILRV competition, NASA and the Air Force had been developing lifting bodies as research craft and space reconnaissance vehicles: throughout 1968 and 1969, a series of lifting bodies underwent piloted drop tests from NB-52 bombers at Edwards. In addition, the Air Force launched into space and recovered several unpiloted lifting bodies, beginning with tests of the X-20 heat shield in its ASSET (whose name stood for "Aerothermodynamic-Elastic Structural Systems Environment Tests") program and continuing through the 1966 test of a miniature lifting body under the X-23 PRIME (which stood for "Precision Recovery Including Maneuvering Entry") program.[75]

Martin Marietta's X-24 lifting body the following year was designed to eventually serve as a piloted spaceplane for NASA and the Air Force: plans to launch a full-scale version atop a Saturn launch vehicle came close to fruition, and a launch aboard an Air Force Titan launch vehicle was the subject of frequent discussion.[76] X-24B drop-test flights at Edwards continued through 1975, and while Lockheed Martin (the two manufacturers merged in 1965) canceled production of the X-24C in 1977, analysts have speculated that a full-scale Air Force lifting body reached space either during the 1970s, or allegedly later, as part of the classified Blackstar program.[77] Decades later, the Air Force flew its unpiloted, winged, reusable X-37B shuttle, but historians' understanding of the shuttle mode debates of the 1960s remain clouded in a lack of available records on classified programs of the era.

Between 1968 and 1971, the space shuttle's likely configuration remained unclear. Most designs for winged spaceplanes (and criticisms of them) were not new: the stumbling blocks that had plagued earlier design competitions remained, as no new technologies had yet emerged to obviate them. Major questions concerned the size of the orbiter and whether it would have its own engines or serve as a mere "glider" propelled by boosters, a question that bore greatly on the cost of the system. Also considered was an inexpensive, expendable fuel tank affixed to the reusable engines mounted on the orbiter or a separate booster stage, which could be recovered and refurbished for each flight, creating a technically simpler booster than the piloted fly-back version, but one that would fall short of true reusability.[78] Somewhere lower on the list, and still repugnant to NASA, was the idea of using an expendable launch vehicle (like the Air Force's Convair SM-65 Atlas ICBM) to launch a small, piloted glider, sacrificing reusability and cargo capacity in order to sustain the bare premise of a shuttle. Even Darrell Romick advocated for this cost-effective approach, though, offering a scaled-down "METEOR, Jr." that "could even be launched by an expendable Atlas-sized vehicle!"[79]

Still lingering in the background, almost unstated and practically forbidden to discuss, was the option of developing a new generation of expendable boosters and piloted capsules, which would be easy to design but expensive, NASA feared, to operate in quantity. More important, such vehicles threatened to undermine the mythology used to justify the entire shuttle program: a cheap vehicle had to be reusable, and the only vehicles that were truly reusable were those that could land on runways. This already chaotic and bewildering contracting process was rendered even more so by the fact that some contractor proposals were secretly produced by moonlighting NASA engineers hoping to steer the selection toward North American Rockwell, their preferred partner.[80]

As specifications and capabilities continued to change in NASA's contractor guidance, increasingly the only system feature that appeared to stay constant was the necessity that the craft land on runways, a requirement near to the heart of many NASA managers but one with few direct benefits.[81] In addition to rejecting scramjets, Mueller, for example, rejected any transportation system featuring either expendable rocket boosters or a crew vehicle that could not achieve a runway landing. Meanwhile, proposals for gliding space shuttles that strayed too far afield—like those utilizing lifting bodies instead of conventional wings—NASA eventually dismissed, ostensibly for technical reasons. Testing revealed that the mass efficiency and reentry performance of lifting bodies were often exchanged for poor handling upon landing, as Eugen Sänger had discovered three decades earlier.[82] Although theoretically solving certain problems related to frictional heating on reentry, NASA also feared the "squashed-lightbulb" shape of lifting bodies would be difficult to fabricate and fill with subsystems and cargo modules, which were apt to be cylindrical or cube-shaped.[83]

These debates were among the more visible of the shuttle design controversy, as each fundamentally changed the appearance of the craft intended to embody the American future in space. Ultimately, each of the approaches required the solution to additional, less publicized problems that had vexed spaceplane designers for decades. More pressing than any of the booster debates, for example, was the search for a reliable means of protecting the shuttle from frictional heating, using a material that was mechanically durable, reusable, and light enough not to weigh down the whole craft so much that no rocket could launch it.

This search was largely invisible to the American public but had proved the most worrisome in previous iterations of the spaceplane concept, and, at the conclusion of the space shuttle program in 2011, remained imperfectly solved. In the 1950s, no good solutions presented themselves to the problem of the protection of spaceplanes during their fiery, high-speed return from space through Earth's atmosphere. Most reentry vehicles of the 1950s and 1960s employed ablative heatshields composed of phenolic resins that that carried heat away from them as they charred, but the shields decomposed with each use. Researchers looking for a reusable heatshield fixed upon a variety of reusable heat-management technologies, most of which traded temperature resistance for weight or physical vulnerability. Metal alloys that could withstand high heat oxidized readily in the atmosphere, requiring delicate coatings to maintain their chemical integrity. Metal heat sinks, which absorbed excess heat and stored it in other parts of the vehicle, added significant weight to spacecraft, as did active refrigeration systems that cooled the vehicle's skin with pipes full of water. Ceramic firebricks made mostly of sand (like those used to insulate water boilers) offered a technically simple means of insulating a metallic fuselage, but they were heavy and lacked the flexibility and tensile strength of metal. If used to cover a spaceplane, they would likely crack and shatter well before reentry.[84]

These technical difficulties were impossible to conceal from experts. Although the shuttle enjoyed critical support inside NASA among its senior management, particularly administrators James Webb and Thomas Paine and associate administrator Mueller, support from scientific and technical experts outside NASA was thin. By 1969, even NASA's in-house experts balked at the Frankenstein's monster that the shuttle was becoming, and contractor proposals slowly accreted features while fundamental problems remained unsolved. Low had stepped away from shuttle development for "several months" in mid-1969, only to return to it in August to find himself "speechless" at what NASA had decided in his absence. In an August 5 memo to Robert Gilruth, Low detailed a series of concerns about the shuttle's growing list of odd requirements, which pleased enthusiasts for the system but imposed considerable design challenges and offered little economic benefit. Among these were the notion that the shuttle, when on Earth, could ferry itself around the country using its own fuel tanks and air-breathing jet engines, and that simultaneously, it would have sufficient wing size to land in the same location from which it launched (obviating the needed for "self-ferry capability") (figure 2.4). More critically, Low feared that the process by which

Figure 2.4
NASA's preferred configuration for the shuttle orbiter included jet engines that would enable the craft to cruise in the atmosphere for a considerable distance and enable it to land at airfields, as in this concept drawing by North American Rockwell. NASA photo, courtesy of NARA, National Archives Identifier: 17409730.

such features had found their way into shuttle requirements suggested that NASA's regular procedures for approving changes to human spacecraft were not being followed. "Have we really performed systems analyses to demonstrate these points," Low mused, "or are they merely wishful thinking?"[85]

Apollo, Low noted, had experienced a troubled development until NASA instituted Change Control Boards staffed by experienced managers to verify that new additions to a spacecraft's mission would not cause a cascading series of complications. "I am not aware of any forum where people like [Apollo Flight Director Christopher] Kraft, [Chief Astronaut Deke] Slayton, and I have been involved in decisions leading to these ground rules," Low noted with displeasure.[86] Low, in the memorandum and accompanying transmittal note, demanded that NASA form a shuttle Change Control Board "so that people like Chris Kraft and others with some experience in the business can have a major voice in the future direction of manned flight."[87] Otherwise, Low feared, NASA would be "starting out making the same mistakes that we made 9 years ago when we issued the first Apollo statement of work."[88]

Conclusion

While NASA facilities, defense contractors, and consultants fine-tuned proposals to meet NASA's needs, oversight agencies like the OMB weighed-in on potential designs, offering recommendations that occasionally threatened the survival of the STS. The shuttle's multiple identities—a human spacecraft, a carrier for scientific satellites, an orbital laboratory, a craft to service space stations and interplanetary ships, a delivery vehicle for commercial payloads, a tool of diplomacy and international cooperation, and, eventually, a military spaceplane—were always in conflict, and as each required a different machine, NASA could never satisfactorily reconcile them. Within a few short years, the shuttle had transformed from an optimistic effort to create a spacefaring civilization to an awkward, winged vehicle that satisfied no one, including Air Force planners whose interest in the shuttle increasingly determined its design.

3 The Winged Gospel and National Security, 1969–1974

In *The Winged Gospel: America's Romance with Aviation, 1900–1950,* Joseph Corn (1983) writes that while Americans gravitated toward technologies of flight in the early twentieth century, the same enthusiasm did not always extend to the space vehicles of the later decades of the century. The airplane had been a revelation and a miracle, Corn notes, in part because people regarded it as a technology in which all would partake. Journalism, barnstorming, public display, and hobbyist activities in the early decades of the twentieth century created an attitude of "airmindedness," defined principally by a conviction that "soon everybody would fly," whether in an airplane of their own or through some form of "aerial mass transit."[1] Later space programs like Project Apollo, though, were demonstrative rather than participatory: average people could watch spaceflight on television but could not participate in it.[2]

To its proponents at the National Aeronautics and Space Administration (NASA) and private industry in the mid-1960s, a winged space shuttle—familiar, safe, and inexpensive to fly—promised to bring airmindedness into the twenty-first century by creating a comfortable spaceflight experience even for the lay public. While some imagined the spaceplane as a scientific tool, others saw a tourism ship designed for amateurs and adventurers or an instrument of international diplomacy, much as early futurists had envisioned the airplane.[3] These visions, though, were shared by relatively few people. For most of its early history, the space shuttle was an orphan project, popular with certain engineers but not with the politicians who hired them.

The space shuttle's transformation from a dream of the future to a funded vehicle between 1968 and 1972 owed its existence principally to domestic politics, national security concerns, and the ascendance of a new "winged gospel," in which vehicles with wings were increasingly seen as the only proper spacecraft. NASA had previously examined dozens of potential spacecraft designs (ballistic capsules, lifting bodies, and winged gliders), unable to decide among them. Within a few short years, though, the idea that the US would build a spacecraft that did not resemble a traditional airplane became unthinkable. This transformation was driven by NASA's insistence that "spacemindedness" required such a vehicle and was cemented by the needs of a critical new partner in the shuttle's development. Solicited by NASA to help sell the shuttle to the executive branch, the US air force forced a change in the shuttle's size and shape, pushing it closer to the design of a conventional airplane. Rescued by politics and reshaped by compromise, the shuttle emerged as a product of forced collaboration, its purpose and design muddled to secure approval from a presidential administration never truly convinced of its value.

"Not One of Those Space Cadets"

Assuming the presidency in the winter of 1969, Richard Nixon began his first term in office almost as skeptical about civilian spaceflight as the previous Republican president (and Nixon's former running mate), Dwight Eisenhower, had been. Eisenhower dismissed the Soviet Union's launch of Sputnik 1 in 1957 as a "stunt,"[4] but Nixon had been appreciative of spaceflight's potential public relations impact. This enthusiasm, however, did not translate, during the first months of Nixon's own presidency, into a desire for large expenditures on civilian space exploration.[5] At times eagerly chatting up astronauts and determined to see the US triumph in space, Nixon was just as often openly contemptuous of spaceflight, concerned only with how success or failure would affect his reelection campaign.[6] ("I am not one of those space cadets," Nixon confessed to a congressional liaison in March 1971.)[7] Not surprisingly, the decision to develop the space shuttle, as the historian and Columbia accident investigation board member John Logsdon noted, was chaotic from the outset. Nixon's ambivalence on the subject and the inexperience and the venality of his White House advisors combined to muddle debates on space policy and produce outcomes that satisfied nobody.[8]

With few persistent opinions about space policy, Nixon relied heavily on a small collection of White House staff members to make decisions for him; these often-unqualified political operators isolated Nixon from actual experts in the executive branch. When motivated to do so, they usurped decision-making authority themselves, but when confronted by painful choices, they often concealed matters from Nixon and attempted to stall as long as possible.[9] Nixon himself was hapless in managing these men: constantly distracted, he generally agreed with the last person with whom he had spoken, producing an aggressive contest between advisors to control access to him. NASA leadership at the time, though, was also weak and in flux: the Apollo-era administrator James Webb left the agency in October 1968, and NASA proceeded through five more administrators in ten years as it struggled to garner enthusiasm for post-Apollo human space exploration.

From the beginning of Nixon's first term, the experts he convened to weigh in on NASA's space shuttle plans were unsupportive. On January 8, 1969, Nixon's team assembled a thirteen-member study group (led by the Nobel Prize–winning physicist and inventor of the laser, Charles Townes) to advise the incoming administration on space policy; it issued a confidential report recommending against the development of the shuttle without even mentioning it by name. "The unit costs of boosting payloads into space can be substantially reduced," the *Townes Report* wrote, alluding to the space shuttle, "but this requires an increased number of flights, or such an increase coupled with an expensive development program. We do not recommend initiation of such a development." In the report, the Townes Task Force identified the critical problem with early NASA estimates of the shuttle's utility: its cost savings required a more intensive space launch schedule than anyone had contemplated.[10] There was no groundswell of support of the space shuttle, and little enthusiasm for it even among America's scientific and technical elite.

Nixon remained unenthusiastic about the space shuttle for most his first term; three years after assuming the presidency, though, he provided ambivalent approval to begin space shuttle development, an endeavor he confidentially derided as a "stunt," just as Eisenhower had.[11] This transformation owes less to the merits of the shuttle than the competing constituencies that Nixon believed the craft would satisfy, a lingering fear of being blamed by history for abandoning a high-technology project that promised to restore American confidence and global leadership, and elements of naked political calculation. While the Space Task Force described numerous benefits of space exploration—from scientific discovery to improvements in telecommunications—diplomacy, prestige, and national security were still the only ones that it considered important.[12] Were he to abandon spaceflight, Nixon feared a critical loss of his and the nation's international reputation at a time of political, economic, and social unease. Instead, he promoted the craft for its potential to create opportunities for average people to fly in space, the same role that the airplane had provided half-a-century earlier.[13] In private, he also mused about its potential to win him domestic political support.

To pass muster with Nixon's staff, NASA's shuttle needed to provide solutions to problems more practical than the democratization of spaceflight. Effectively redesigned by the Air Force as a reconnaissance aircraft, the shuttle underwent physical transformations between 1969 and 1972 that muddled its design, until reusability and even basic safety were fatally compromised. Throughout its development, only a single feature of the space shuttle appeared to be nonnegotiable: its wings. The original shuttle design that NASA's Langley Research Center had envisioned would have been a fully reusable, winged aircraft, with a piloted, reusable winged booster. The winged booster soon fell by the wayside, and the shuttle's ultimate design was instead an amalgamation of reusable and expendable parts that satisfied neither fiscal conservatives nor space enthusiasts, but did retain the orbiter's wings. Participation by the Air Force in shuttle design decisions, as Stephen Garber and other scholars have written,[14] ensured that the craft would keep this configuration, and thus the winged space shuttle did not represent the solution to a technical problem so much as the answer to a cultural and political one.

Impossible to develop on the budget allotted to it, NASA's space shuttle underwent a series of capability-reducing restrictions between 1969 and 1974, until it looked nothing like what its inventors had imagined.[15] To the historian Howard McCurdy, debates surround the shuttle's mode of operation constituted NASA's first true "political baptism": in the absence of the kind of consensus that animated Project Apollo, NASA had to lobby various groups to achieve its goals. No single shuttle concept fulfilled all these diverse objectives, so the question became not how NASA would succeed, but in what ways it would prefer to fail. Wrote McCurdy: "Unable to get their overall vision approved, NASA scientists and engineers plunged into the morass of incremental politics. They had to negotiate shuttle design details with White House staff. They felt obliged to accept a technologically inferior program in order to win political support, and they had to engage in the game of bureaucratic politics, seeking outside support from groups like the military, who came to NASA's aid."[16]

Rather than negligent oversights, the design compromises that came to define NASA's space shuttle were purposeful accommodations to political and cultural goals, known to impose increased risks to the system. And instead of ameliorating these risks immediately, NASA hoped that additional funds that Congress appropriated in the future (or that NASA earned through commercial operations) would permit a redesign of the shuttle once it had proved to be practicable. Before that could happen, though, NASA negotiated a series of design modifications that were necessary to secure political support for the craft, but that all but doomed it to operational disaster. In fact, NASA eventually sacrificed even reusability to maintain the shuttle's winged, piloted character, a fateful decision ultimately responsible for the critical failures of the Space Transportation System (STS).

Enter the Air Force

Throughout the early months of Nixon's first term, the space shuttle lay largely dormant as a national priority: although the concept had many fans within NASA, it enjoyed few friends outside the agency and its contractors, even among spaceflight advocates. Enthusiastic instead about the prospect of a piloted flight to Mars, Vice President Spiro Agnew, chairing Nixon's Space Task Group, advocated strongly for Mars flight as NASA's next large-scale project, pushing the reusable, winged Earth orbiter into the background. Rather than advancing NASA's interests, the agency feared that Agnew's advocacy was counterproductive: neither Nixon, Congress, the media, nor the general public favored the expenditure of funds for a human landing on Mars,[17] and single-minded advocacy of such a scheme would likely irritate Congress and doom NASA to having no Project Apollo follow-on program at all. The space shuttle, by contrast, appeared to offer more practical benefits and enjoyed enthusiastic support among key NASA managers under Administrator Thomas Paine, including George Mueller and Wernher von Braun, who secretly hoped not only to build part of the craft, but fly it into space himself.[18]

Von Braun was a particularly forceful advocate for the craft. Recounted NASA deputy administrator Hans Mark:

> I have to confess that I was not an enthusiastic supporter of the space shuttle program at the time. My own background in space science had conditioned me to think in terms of unmanned space operations and I felt that the presence of people might actually complicate things. . . . I also felt that the shuttle then being

> considered was too large and that it would be best to evolve a space shuttle from a series of smaller vehicles much along the same lines as the Air Force's recently cancelled Dyna-Soar Program. . . .
>
> It was von Braun as much as anyone who persuaded me that the space shuttle . . . was indeed the right first step.[19]

Mark recalled that von Braun's advocacy hinged upon two arguments: a human presence in orbit would augment NASA's space science efforts (an argument in favor of human spaceflight, but not necessarily for a space shuttle), and the shuttle's introduction would hasten the construction of a permanent space station, which Mark and others within NASA strongly supported. The latter argument, though, was far from clear: von Braun asserted that the shuttle was harder to design and build than a space station, necessitating that the shuttle be built first to enable the latter's construction.[20] A space station could also be built—though more expensively and slowly—using expendable launch vehicles, so the timetable of space station construction thus seemed to be the same whether NASA built the shuttle or not. Von Braun's case for the shuttle was a tautology: the shuttle needed to be built because if it were not, there would be no shuttle.

Despite von Braun's enthusiasm, the shuttle was in a precarious political position in 1969: Mark remained an "inside critic," along with others in NASA, and support for the shuttle outside of NASA was virtually nonexistent.[21] The Air Force, whose own spaceplane plans had been repeatedly thwarted was, as Mark surmised, an obvious partner whose enthusiasm for piloted spacecraft equaled von Braun's. Nixon's Space Task Group agreed, encouraging such cooperation in March 1969, shortly after Nixon took office.[22] With Mars flight unlikely to be funded, Paine reached out to the secretary of the air force (and former NASA associate administrator), Robert Seamans, in April 1969, proposing joint study of the shuttle to "meet the needs of both agencies."[23] Seamans was not convinced: he voiced his skepticism about the shuttle in NASA's August 4 briefing about post-Apollo space plans, concerned that the technology to build the craft was not yet available.[24] Eventually, Seamans offered only a qualified endorsement, agreeing with NASA that "the extent to which space is used either for exploration or national security depends upon the cost per pound in orbit," without committing to any particular architecture.[25]

Meanwhile, NASA continued to promote the idea that the shuttle was the centerpiece of a broad program of civilian spaceflight, though it placed most of the goals for planetary exploration well into the future. On September 11, 1969, Mueller wrote to NASA Kennedy Space Center director Kurt Debus to explain the proposed schedule of (as yet unfunded) explorations, beginning with those already in development, like the "Saturn V launched Workshop," later named "Skylab":

1972	Apollo Applications Program (AAP) operations using a Saturn V launched Workshop
1973	Start of Post Apollo lunar exploration
1974	Start sub-orbital flight tests of Earth to orbit shuttle
	Launch a second Saturn V Workshop
1975	Initial space station operations
	Orbital shuttle flights
1976	Lunar orbit station
	Full shuttle operations
1977	Nuclear stage flight test
1978	Nuclear shuttle operations—orbit to orbit
1979	Space station synchronous orbit
By 1990	Earth orbit space base
	Lunar surface base
	Possible Mars landing[26]

A critical decision in January 1970 placed the shuttle first in NASA's list for future projects, unseating even the planned large orbiting space station that it would ostensibly service.[27] Increasingly, it appeared that NASA would develop the shuttle no matter what, but the Air Force would have the opportunity to join in the program on the ground floor, ensuring its use of the vehicle throughout the life of the project.

The difficulties associated with this a partnership became apparent almost immediately. Discussing their collaboration on January 27, 1970, Seamans and Low addressed a variety of problems suggesting that the goals of the NASA and the US Department of Defense were simply incompatible, but throughout the discussion (as Low later recounted), it was NASA rather than the Air Force that appeared to yield. Seamans had insisted that NASA build a larger vehicle than it intended, and one with larger wings. Low agreed verbally to most of the Air Force's design demands, and when Low raised objections to some of them, he took Seaman's silence as agreement with Low's concerns rather than a dismissal of them. Seamans, as Low later noted, also "mentioned that, in the 1960's [sic], NASA was fully supported because of the competition with Soviet Russia," and that "this type of support should not be expected in the 1970's [sic]."[28] So desperate was Low for Air Force support that he accepted Seamans's threat without complaint: NASA would be lucky to get any support from the Air Force, and the Air Force might walk away at any time just the same.

Although Seamans sympathetically acknowledged Low's concern that the shuttle's proposed military and intelligence work would complicate NASA's identity as a civilian scientific agency (and damage NASA's international relations), he did little to assuage Low's fears. Low seemed satisfied merely to have his concerns heard. As a negotiation, it was a staggering victory for the Air Force: Seamans would determine the size and shape of the shuttle, receive operational parity—even priority—in using the vehicle, tarnish NASA's global image, and pay nothing for the privilege. In return, NASA would receive only the Air Force's assent in pitching the program to Nixon's staff of skeptics. With the Air Force easily able to scuttle NASA's shuttle ambitions, the deal was borderline extortion, with the Air Force effectively threatening to torpedo the shuttle with Nixon unless NASA agreed to reconfigure it as a military program to support space-based reconnaissance.

Spy satellites had been under active development in the US since 1946, when the RAND Corporation published the first in a series of reports outlining the parameters for viable space-based reconnaissance platforms. Such satellites, the reports indicated, should be launched southward into orbits that would circle the Earth in ninety minutes. As the Earth rotated beneath the craft, the satellite would cross the equator in successive tracks 1,000 miles apart. Delivering reconnaissance equipment into orbit using larger, piloted spacecraft like Air Force Dyna-Soar at first appeared attractive but posed unnecessary complications. Such a craft might remain in orbit for hours (or at least until it flew over a suitable runway), and once in orbit, it would likely be observable from the ground after a few cycles. Simple calculations would enable a determination of the spy vehicles' orbital parameters and flyover schedule. Beginning in 1959, the Air Force, Navy, Central Intelligence Agency, and National Reconnaissance Office (which, in 1960, assumed responsibility for spy satellites) employed a series of expendable Air Force rockets to deliver unpiloted photographic reconnaissance and radio eavesdropping satellites into orbit. The depleted rocket boosters that carried them reentered the Earth's atmosphere and disintegrated minutes after launch, partially concealing the satellites' mission.

With the space shuttle, the Air Force resurrected the piloted reconnaissance vehicle concept in the hope that improved design would alleviate Dyna-Soar's deficiencies. To be most effective as a reconnaissance platform, a piloted spacecraft would need to return to Earth after a single orbit, but as the Earth would rotate beneath the orbiting vehicle, by the time it returned to Earth, its launch site would be over 1,000 miles away. Chief among the Air Force's design requirements for the shuttle was wings large enough to enable the shuttle to glide far enough during reentry and landing to return to its launch site or travel to a second designated location.[29] The purpose of the shuttle's proposed (and enormous) 1,500-mile cross-range capability, was, almost absurdly, to enable astronauts to launch from California, observe geopolitical events from space, and then either return to their launch site or land near Washington, D.C., to brief their superiors.[30]

The requirements for cargo capacity (i.e., large enough to hold a telescopic camera the size and shape of a school bus) and gliding performance all but demanded an orbiter with conventional wings, effectively ending any real debate on the shuttle's design. Although Logsdon speculated that NASA may have chosen the same design even absent Air Force pressure,[31] the NASA engineer R. Dale Reed later concluded that the requirement to carry large reconnaissance satellites severely constrained NASA's choice of the configuration of the vehicle, which now had to be built around its intended cargo rather than the other way around:

> The payload requirement of the Air Force was about 50,000 pounds to low orbit, to be contained in a compartment roughly 15 by 60 feet, or about the size of a railroad boxcar. Easy access to this compartment also required the use of full-size doors that could be opened in space. This requirement narrowed down the potential spacecraft shape to what basically resembled a rectangular box with lifting surfaces (wings and tails) attached to it, plus a rounded nose on the front and rocket motors on the back.[32]

Adding spying capabilities and cross-range to the shuttle not only required larger wings,[33] it multiplied the thermal protection problems for the craft and increased its weight, which in turn restricted the options for an appropriate booster.[34]

While demanding enhanced capabilities, Air Force planners also insisted on reduced-cost construction, further burdening contractors attempting to design a craft that could operate safely. To conserve precious supplies of heat-resistant titanium needed for other aerospace applications,[35] the Air Force required that the shuttle's fuselage be manufactured from cheaper and more abundant aircraft-grade aluminum. A shuttle made from aluminum could be fabricated at considerably less expense than one made from titanium but required an even more robust thermal protection system (TPS). Metallic heat shields produced from titanium alloys offered the best protection from impact as well as heat, but their use on an aluminum fuselage was impossible. The juxtaposition of dissimilar metals in construction would produce oxidation that would rot the craft from within, requiring the use of one of the more physically delicate, nonmetallic heat shield technologies instead. A large, winged shuttle made of aluminum could fly, but it posed significant additional design challenges.

These design debates proved so vexing that NASA and other government agencies considered tabling these questions until a later point, when the constituent technologies had matured. Through 1971, it appeared increasingly likely that production of a fully reusable space shuttle would be phased in over many years, with NASA first commissioning a shuttle and booster built with single-use parts.[36] Suggestions included crafting an expendable winged shuttle from existing Saturn hardware: von Braun continued to suggest that NASA use the Saturn V to boost a prototype shuttle orbiter into space until the winged, reusable booster became available. Low was also an early supporter of launching a prototype orbiter atop an expendable Saturn V S-IC first stage, and further debate through the summer of 1971 suggested that the STS would probably need to progress through a series of intermediate versions as funds became available to complete it.[37] Despite public statements to the contrary, senior NASA managers recognized that a winged space vehicle need not be reusable, and even a reusable shuttle would not need a reusable TPS: ablative heat shields that were consumed during reentry could be replaced between flights, offering NASA a cheaper vehicle that it could upgrade later with a reusable TPS. So, too, for the shuttle's reusable main engines: NASA could fly a prototype shuttle with expendable engines ("Mark I") until new, reusable engines qualified for flight ("Mark II"). Budgetary limitations imposed by the Nixon administration through 1971 suggested the likelihood of a phased shuttle program whether NASA management wanted one or not, a fact that even von Braun and the newly appointed NASA administrator James Fletcher acknowledged.[38]

Most NASA managers, though, were publicly hostile to the incremental approach regardless of their private support. Arguments for the development of a spaceplane hinged upon the development of a craft that was so cheap to fly that it would create markets for spaceflight that did not yet exist. Efforts to reduce development costs would reduce the reusability features necessary to realize operational savings, while probably increasing the long-term development costs as well. As the NASA acting shuttle program manager Charles Donlan described in a July 1972 retrospective on shuttle development:

> Time phasing some of the orbiter subsystems received considerable study effort. This was known as the Mark I/Mark II shuttle system. The Mark I orbiter was to use available ablative thermal protection, a J-2S engine developed as an extension of the existing Saturn J-2 engine, and other state-of-the-art components such as existing avionics. Improved subsystems such as fully reusable thermal protection and the new high pressure engine would be phased into later orbiters to achieve the operational system (Mark II). This time phasing reduced expenditures early in the development cycle but the Mark I system had reduced payload and crossrange capability as well as an increased turnaround time of one month. This represented a severe loss in operational capability. Furthermore, the total development costs to achieve the full Mark II system actually increased.[39]

NASA's initial rejection of the phased approach reflected its awareness of just how politically precarious the shuttle was; though reasonable as a means to reduce development costs, the phased approach jeopardized the agency's vision for the shuttle and produced internal dissension on system architecture instead of consensus, just when consensus was most needed.[40] Incremental design and testing, Donlan concluded, would not solve the shuttle's basic problem: for the final version of the vehicle to achieve the promised economies of operation, all of its constituent technologies—from reusable engines to its heat shield to its reusable booster—would need to work from the very beginning of flight operations. Phasing development also made Air Force participation unlikely, as only a large, fully realized STS offered cost advantages over the Air Force's existing fleet of expendable launch vehicles.[41]

Furthermore, if NASA chose a phased development approach, it might encounter two unpleasant alternatives to the completion of a reusable shuttle. Even if modestly successful, a nonreusable Mark I shuttle was a costly, short-term experiment, more likely to be canceled than a fully functioning orbiter would be. And if the nonreusable Mark I shuttle worked well, it might never receive the additional funding required to make it reusable because the Mark II shuttle would be extremely expensive to develop, and the argument in favor of full reusability had never been very strong to begin with. Having pitched the shuttle as an "all or nothing" proposition to gain support for the features that it wanted, NASA was now trapped by its own analysis: anything short of a full-sized orbiter offering full reusability would torpedo NASA's tentative mandate to proceed.

NASA formalized its collaboration with the Air Force in February 1971, emphasizing the shuttle as "an economical capability for delivering payloads of men, equipment, supplies, and other spacecraft to and from space by reducing operating costs an order of magnitude below those of present systems."[42] The Air Force's lukewarm endorsement was enough to tip the White House scales on the shuttle decision, but at first, NASA and the Air Force struggled to identify a viable military purpose for the craft, just as Robert McNamara had wrestled with Dyna-Soar. Attempting to justify shuttle development in a 1971 memorandum to the president, Fletcher admitted that "our military planning has not yet defined a specific need for man in space for military purposes." Unwilling to concede the overall point, though, Fletcher continued, asking, "Will this always be the case? Have the Russians made the same decision?" Even if the Air Force required no piloted space vehicles, Fletcher intimated, the shuttle would, at the very least, provide a "quick reaction time and the ability to fly ad hoc military missions whenever they are necessary," and "even without new military needs, the shuttle will provide the transportation for today's rocket-launched military spacecraft at substantially reduced cost."[43] The promise was striking, suggesting that despite the shuttle's status as a civilian space vehicle, military payloads had priority over shuttle cargo manifests. The statement also suggested that the shuttle would be configured principally to handle military satellite traffic: a significant engineering concession that threatened to undermine its safety and economic viability.

Although later claiming to have no preference for any proposed shuttle architecture, Fletcher appeared to recognize that a large, winged craft meeting Air Force requirements was the only vehicle that Nixon would fund.[44] This was a poor argument for its construction, however, leaving NASA vulnerable to criticism. The critiques that the rocket pioneer and Navy captain Robert Truax leveled at early designs for what became NASA's space shuttle anticipated the negative analyses that bedeviled it for the rest of its career. An engineer who had begun his research with experiments conducted in his spare time at the US Naval Academy during the late 1930s and continued through the Air Force's successful Thor program, Truax had become a forceful advocate of large expendable launch vehicles by 1970, finding in the shuttle only a hodgepodge of dated design ideas and superfluous features. Chief among the "frills" that Truax noted was NASA's insistence that the shuttle orbiters possess wings, and that even their boosters have them as well:

> The features which I consider in the category of peripheral frills, but which present vast difficulties, include land touchdown and booster flyback. These features, unfortunately, are near and dear to many proponents of reusable vehicles. They make the "aero" part of the aerospace industry feel needed. They even have appeal to the non-technically minded. But they make about as much sense as requiring airplanes to be able to land at railroad stations.[45]

With no Air Force interests to advance, Truax saw the space shuttle's design as evidence of misplaced priorities. In the minds of NASA and contractor engineers, runway landing appeared to take precedence over every other aspect of the shuttle's design, including reusability, which was the true path to realizing cost savings from the system. A reusable craft, though, did not need to have wings and would be cheaper to build without them. By pinning its hopes on a space vehicle designed to be as airplanelike as possible, Truax argued, NASA had given up on doing something truly revolutionary after Project Apollo. An enthusiast for building an even larger rocket than NASA's 364-foot Saturn V—the Sea Dragon—Truax saw in the space shuttle not an impressive new vehicle for deep space exploration, but rather a utility craft sacrificing performance for "elegance," and accomplishing, in a more challenging way, what disposable launch vehicles were already doing.[46]

Truax implored Fletcher to reconsider the shuttle program for this reason in personal correspondence to him dated June 22, 1971, and received by NASA three days later. Rather than dismissing reusable space vehicles completely, Truax reminded Fletcher that Truax had worked actively on the subject since retiring from the Navy in 1959, and he proposed reusable space vehicles for the American Institute of Aeronautics and Astronautics in 1966, provided that they were simple in design. At the time, Truax noted, NASA had rejected these modest proposals, insisting that "foreseeable space traffic (including a projected Mars mission) was insufficient to amortize the development of a reusable vehicle."[47] Yet, despite any fundamental change in the economics of spaceflight within the last five years, NASA now embraced the reusable space shuttle wholeheartedly "for some reason." Truax feared the decision to build a "complex and difficult" craft like the shuttle contained the seeds of personal disaster for Fletcher and would "cripple the space program for years to come."[48]

Reusability, Truax intimated, had never driven NASA's enthusiasm for the shuttle; instead, it was the product of a pathological fascination with winged spaceflight among aeronautical engineers. Repeating his public criticism, Truax continued that the goal of reusability had already been met with nonwinged craft: NASA and Air Force rockets like Aerobee and Titan III had been successfully recovered at sea intact and even reused, obviating the need for "unnecessary claptrap" like wings and jet engines.[49] Predicting that NASA would experience "the biggest fiasco in its history," Truax implored Fletcher to distance himself from the "great space-plane flop" and avoid wasting taxpayer funds on a program with such small chance of success, especially when cheaper alternatives were readily available. He continued, "It would be especially the pity since we can meet 90% of the real goals at 20% of the cost without really stretching the state of the art one bit."[50]

A respected naval officer and rocket engineer with a connection to the NASA engine supplier Aerojet, Truax was an individual whom Fletcher had to take seriously, even if his pointed criticisms were unwelcome; Fletcher instructed the associate administrator for manned space flight, Dale Myers, to ghostwrite a technical response to Truax, which Fletcher would "rework" to make it "more personal."[51] Critically, Myers did not seek to counter Truax's technical or economic arguments; instead, he claimed merely that NASA and its contractors were in the process of developing new engines and materials that would reduce the weight and cost of the shuttle until it made economic sense. These responses hinted at the incremental development approach that NASA had already rejected, but they were the only way to deflect Truax's trenchant criticisms. Rather than declaring the technical and economic problems of the shuttle solved, Fletcher indicated that the shuttle's design remained uncertain, and promised that further studies eventually would produce a craft that even Truax would support.[52] The curt response was a dismissal rather than a direct reply to Truax's critique and offer of alternatives; NASA would produce the shuttle and had no interest in other architectures.

Myers had been an interesting choice for Fletcher to press into service in the defense of the shuttle. A year-and-a-half earlier, Myers had received similar criticisms, this time from Low, whose August 1969 concerns about the shuttle had only intensified in the months that followed. On January 27, 1970, Low had written to Myers that NASA and its contractors had been confusing the shuttle's "objectives" ("to provide a low-cost, economical space transportation system") with its "requirements": a loose set of features like "cross-range, go-around capability, fly-back capability, and even payload size and weight," which were largely unnecessary and produced greater development costs.[53] Although at first seeming to argue for the abandonment of the entire shuttle effort in favorable of a disposable launch vehicle and capsule based on the Air Force's Titan III rocket and NASA's Gemini spacecraft, Low concluded by suggesting that the shuttle program could continue, but only if costs could be lowered so much that the shuttle "creates its own payload market."[54]

By summer 1971, the Air Force, never a strong proponent of the STS appeared increasingly uncommitted to it, leading NASA to conclude that its erstwhile partner was no partner at all. NASA insiders briefed on the agency's collaboration with the Air Force were internally skeptical that the Defense Department would honor its vague promises to fly all military payloads on the shuttle. Fletcher feared that even those engineers who were ardent shuttle supporters—people convinced of the shuttle's "technical feasibility" and "cost effectiveness"—doubted that military users would "agree to give up the boosters" they were planning to field during the shuttle's years of operation. Indeed, experts had warned Fletcher that "the Air Force will not, when the time comes, really give up their other launch capability." Galling to the Air Force, meanwhile, was the fact that so many NASA engineers were themselves ambivalent about the new craft. NASA offices responsible for space station and deep space probe development were also hedging their bets, Fletcher noted, unwilling to remove expendable launch vehicles from their launch plans: "I have had some indications from our own science and applications people that NASA indeed is not even ready to make that adjustment yet."[55]

Ironically, the more NASA contorted the shuttle to fit Air Force needs, the less comfortable the Defense Department became with the alliance; the Air Force was willing to use a NASA spacecraft but did not want to be blamed for its likely failure. US deputy secretary of defense (and Hewlett Packard cofounder) Dave Packard shared such concerns with Fletcher in October 1971, who passed them to Low. Most critically, Fletcher noted, Packard "felt very uneasy about the requirements that had been laid down, especially for the shuttle's cross-range, and payload size," both of which added costs to the shuttle but were driven by Air Force needs. Fletcher was unable to ameliorate Packard's concerns, noting that "the diameter requirement came primarily from NASA and not from the Air Force, but that the length probably came from the Air Force," likely a reference to classified reconnaissance satellites that NASA had agreed to launch for the Air Force in exchange for political support for the shuttle. Packard, Fletcher noted, "knew quite well which program caused the length difficulty," even though it couldn't "be discussed in this memorandum," and he feared "that the payload requirement was somewhat arbitrary at this point."[56]

NASA needed Air Force participation in the shuttle program and made little effort to conceal it; outside NASA and the Nixon administration, noted space program critics like Ralph Lapp and the former NASA scientist-astronaut Brian O'Leary joined in with public criticisms, unconcerned by diplomatic niceties or personal relationships. These criticisms ranged from concerns about the shuttle's military uses to NASA's design compromises to critiques of the various cost analyses (particularly the econometric analyses completed by the consulting firm Mathematica)[57] that relied upon overly sanguine estimates of the shuttle's commercial satellite launch, retrieval, and repair business.[58] NASA projected that commercial entities and foreign governments that used these services would reimburse NASA for them, subsidizing the government's own launch operations and turning the shuttle into a moneymaking enterprise. Rather than lowering the cost of launch services, though, O'Leary feared that the shuttle would soak up NASA funding that was badly needed for planetary exploration to build a spy satellite delivery system that would never make enough money from corporate users to defray its development costs.[59] With estimates of its economics driven by unrealistic assumptions—for example, that it was cheaper to repair a malfunctioning satellite than simply replace it—Congress was being manipulated into funding a system that could never yield sizable cost reductions, and instead would transfer civilian science funding to the military.[60]

Commentators sometimes say of Cold War–era nuclear weapons programs and other products of the so-called military-industrial complex that the military did not corrupt the scientific community; rather, scientists corrupted the military, with promises of weapons technologies that seemed almost too good to be true. Successes in the development of radar and nuclear weapons made virtually any advanced technology seem plausible, and the nation's civilian political leadership proved all too willing to invest in expensive new programs to please voters or generate jobs in swing districts. The shuttle's development, ultimately, was not the case of a unitary military establishment imposing its will on the civilian scientific community. Even within the Air Force (the service branch in which most proponents of spaceplanes could be found), many were agnostic about NASA's proposed shuttle, offering only tepid support and no development funds. Instead, civilians used the Air Force's assent to push a program that they wanted for other reasons: institutional survival (in the case of NASA) and electoral politics (for the Nixon administration).

Sacrificing Reusability

Confronted with a choice of building no shuttle, building a vastly less capable shuttle, or phasing the development of a large and reusable shuttle over many years, NASA ultimately chose the last option, coaxing a mercurial and distracted Nixon into accepting a new version of the craft. By late 1971, the Thrust-Assisted Orbiter Shuttle (TAOS) began to gain enthusiasts in NASA and elsewhere, offering a shuttle as large as the one that the Air Force wanted, but at a vastly reduced cost, achieved by reducing its reusability and physical robustness.[61] Although merely an interim vehicle, NASA's TAOS shuttle would employ a very large, winged orbiter launched into orbit using unpiloted booster rockets, along with reusable engines fed by an expendable propellant tank jettisoned shortly before the shuttle reached orbit. Exactly who first thought of the TAOS configuration remains unclear, especially as antecedents had occurred periodically throughout the Integral Launch and Re-entry Vehicle (ILRV) debates. Mathematica, which had labored under government contracts since 1970 to provide an economic justification for the shuttle, found the most promise in this configuration: Mathematica's Klaus Heiss coined the term "TAOS," and he and his superior, Oscar Morganstern, wrote one of the most compelling arguments for it. Contractors, though, had been mulling over the design for some time, and NASA's Charlie Donlan credited Reinald Finke and George Brady at the Institute for Defense Analyses, an Alexandria, Virginia–based think tank, for the final iteration of the concept.[62]

When it came to the problem of reusability, TAOS was the appearance of a solution rather than an actual one: the new configuration provided a marginally functional flight vehicle large enough for Air Force use until NASA either abandoned the system or found the funds to improve it. The configuration's chief advantage was its low development cost, purchased at the price of higher operating costs: cheaper to build, but more expensive to fly.

Enough like a reusable vehicle to avoid the specter of a phased development, but imperfect enough to justify its eventual replacement, TAOS, as Heiss and Morgenstern would write to Fletcher in an October 1971 memorandum, "abolishes completely the immediate need to decide on a reusable booster and allows postponement of that decision without blocking later transition to that system if still desired." Delaying the transition to a fully reusable system would save money when money was scarce, lowering "the risk and potential cost overruns in booster development."[63]

As TAOS would not be as economical to operate as a fully reusable design, NASA had to work particularly hard to justify its short-term viability. Supporting these hopes was a succession of questionable cost analyses that vastly overestimated the market for the shuttle's services and underestimated operating costs.[64] Mathematica's endorsement of TAOS hinged upon wildly optimistic projections of costs and launch rates; TAOS would make money only if it flew six times as often as any previous spacecraft system, while costing significantly less to operate than other contemporary analyses had demonstrated. Its January 1972 report summarized these conclusions as follows: "WE CONCLUDE THAT THE DEVELOPMENT OF A TAOS SPACE SHUTTLE SYSTEM IS ECONOMICALLY JUSTIFIED, within a level of space activities between 300 and 360 Shuttle flights in the 1979–1990 period, or about 25 to 30 Space Shuttle flights per year, well within the U. S. Space Program including NASA and [Department of Defense]."[65] While TAOS beat other shuttle alternatives in minimizing development costs, the victory was a Pyrrhic one, as profitability required a launch schedule packed with more paying customers than actually existed.

Who these customers were was something of a mystery at the time, but NASA offered predictions. With space stations and interplanetary spaceflight off the table, NASA's justification for the shuttle eventually congealed around the notion that all existing US satellite launch traffic could be replaced by the shuttle, and the shuttle's economy of operation would expand this market even further.[66] A common carrier rather than a research craft, the TAOS shuttle would take on military and scientific missions (sometimes simultaneously), while paying for itself through commercial orbital delivery contracts.[67] These would include civilian satellite deployment, retrieval, and repair, and they might even extend to manufacturing activities, like the exploitation of the weightless environment of orbit for manufacturing flawless crystals, pharmaceuticals, ball bearings, and other sophisticated products for which no existing business existed, but which NASA hoped might be profitable in the future.[68]

By late 1971, even von Braun had turned from a supporter of the shuttle into a near-opponent, fearing that even a cost-reduced TAOS was more than NASA could afford, however much it pleased the agency. (Indeed, the Nixon administration plans to limit NASA's annual budget appeared to make the development of a more robust craft impossible.) After speaking with von Braun in October 1971, Low speculated that von Braun might still support a small glider launched by an expendable booster,[69] but this proposal was more threatening to NASA than any other because the development costs were relatively low and the chances of success great, obviating the need for a more expensive craft. Nixon's Presidential Science Advisory Committee favored this low-cost solution as well: such a craft would satisfy NASA's demands for a winged vehicle and placate the skittish Air Force, which would retain its own space launch capability, independent of NASA programs and schedules. Terrified by what seemed like an entirely reasonable compromise, senior NASA managers struck back, arguing that such a craft, if flown as frequently as NASA hoped to fly the shuttle, would not realize the long-term savings of a larger vehicle.[70] Of course, such a craft would not need to fly as often, and it was this fundamental paradox that revealed the ultimate tautology of the space shuttle. NASA needed to build a large, reusable shuttle because only a large, reusable shuttle could generate the economies of scale needed to justify the expense of building any shuttle at all.

Ultimate responsibility for the decision to approve the TAOS shuttle is difficult to establish: no single meeting produced the decision, and many officials in Nixon's inner circle were disingenuous in representing his opinions to NASA and other agencies.[71] The shuttle itself was an amorphous symbol in larger debates about how the US would maintain aerospace employment and the appearance of a human spaceflight program after the end of Project Apollo. During discussions among key decision-makers, participants often found themselves holding conflicting beliefs as to which variant of the shuttle they were discussing and which individuals were supportive or disdainful. Nixon himself often demonstrated both a hazy grasp of fact and frustrating bouts of ambivalence, changing his mind frequently in response to comments made by different advisors, sometimes within the span of a single conversation. Alternatively admiring and contemptuous of human spaceflight, Nixon wrestled with obsessions shared by nobody else in his administration, insisting, for example, that NASA cancel the final two Apollo flights to the Moon out of a fear that a mishap on either would sink his reelection prospects.[72]

The final shuttle discussions among President Nixon and his White House staff, including Office of Management and Budget (OMB) director George Schultz, Treasury secretary John Connally, Jr., and others took place in November and December 1971; NASA headquarters managers made frequent visits to White House aides during this period, hoping to keep the debate focused on NASA's preferred shuttle configuration.[73] OMB continued to champion the smaller glider, supplemented by a continued Air Force expendable launch vehicle program, to give Nixon both the appearance of a viable shuttle program and a reliable satellite delivery capability for heavy military satellites. Such a compromise would accomplish most of the key goals of TAOS (including the continuation of American human spaceflight and a public relations victory) at greatly reduced cost.[74] A less-capable shuttle design, Logdson later suggested, might have also provided an "evolutionary" approach that would have encouraged NASA to move more quickly beyond the original space shuttle architecture once its high costs became apparent.[75]

Nixon, who was enmeshed in an illegal bribery, burglary, and obstruction of justice conspiracy at the time, appeared distracted throughout the deliberations.[76] His cryptic comments in closed-door meetings confused and alarmed Fletcher, suggesting that Nixon was partial to making the system as cheap as possible—at one point even accepting OMB's recommendation for a vastly smaller shuttle. Indeed, Nixon was taken mostly with the symbolic importance of the shuttle and its guarantee of continued aerospace employment, neither of which demanded any particular spacecraft design. Nixon insisted only that NASA produce *something*, and that, with the 1972 elections quickly approaching, it be manufactured in the electoral swing state of California, a requirement that had no bearing on vehicle configuration (every contractor would gladly produce NASA's preferred design), but all but doomed proposals from the New York–based Grumman Aerospace Corporation, Washington-based Boeing, and Missouri-based McDonnell Douglas (formed in 1967 by the merger of Douglas and the McDonnell Aircraft Corporation). This geographical constraint left only North American Rockwell and Lockheed in the running, with only North American Rockwell (NASA's contractor for Apollo) having experience building human spacecraft.[77] The company and its officers were also major contributors to Nixon's reelection campaign, and Nixon undertook private negotiations with company founder Willard Rockwell, a personal friend.[78]

According to OMB assistant director Donald Rice (who was present at many of the shuttle discussions), NASA's resistance to OMB's smaller shuttle was resolute despite Nixon's enthusiasm for it. Low argued heavily against the smaller orbiter, while Fletcher offered an alternative vehicle of intermediate size, which satisfied no one and may have been part of a cagey negotiating strategy to ensure authorization of the largest option.[79] In the weeks that followed, the ability of TAOS to carry large Air Force satellites carried the day with Nixon's staff of amateur military strategists; while the Air Force preferred to continue using its own expendable vehicles instead of the shuttle, Nixon's staff craved a military justification for the seemingly unnecessary program. NASA's and Nixon's preference for North American Rockwell ensured that it would win the competition to build the final vehicle.

Meeting with Nixon and Fletcher on January 5, 1972, Low later wrote that Nixon, aided by assistant to the president for domestic affairs (and Watergate conspirator) John Erlichman, made a number of candid, contradictory, and often critical statements about the shuttle, suggesting that he was not convinced of its utility but was determined to authorize its construction anyway. Nixon did not appear to have been swayed by any NASA arguments about the shuttle's economics or military utility; he merely wished to be regarded as a spacefaring president and create a positive legacy for his troubled administration. Wrote Low, Nixon "indicated that even if it were not a good investment, we would have to do it anyway, because space flight is here to stay. Men are flying in space now and will continue to fly in space, and we'd be best to be a part of it."[80] Surprising Low, Nixon emphasized only the shuttle's value to aerospace employment and international relations (including possible joint operations with the Soviet Union), dismissing the military uses that his staffers had secretly touted.[81]

Meanwhile, NASA decided on a tentative configuration of the shuttle that moderated some of the worst criticisms of the vehicle, while maintaining the capabilities upon which the Air Force had insisted. The announcement of the awarding of the shuttle's Phase C/D contract on January 5, though, still bore traces of the indecision that had plagued the shuttle debates. Nixon's prepared statement claimed a diverse array of benefits for the new craft, including "international cooperation" and "human betterment," perhaps a sign of the unconvincing economics that would be used to justify its development.

The shuttle, Nixon declared, would simultaneously "take the astronomical costs out of astronautics," and be "safer, and less demanding for the passengers, so that men and women with work to do in space can 'commute' aloft, without having to spend years in training for the skills and rigors of old-style space flight." Critically, NASA offered a new description of the proposed craft as a "good investment" in American spaceflight capabilities, which seemed to acknowledge the lack of a specific purpose for the vehicle.[82]

Fletcher further explained that the reusable, rocket-boosted shuttle would be better able to accommodate diverse activities than previous craft and would promise a more satisfactory work environment for astronauts, all at lower cost. Most significantly, the shuttle was "the only meaningful new manned space program which can be accomplished on a modest budget": "By the end of this decade the nation will have the means of getting men and equipment to and from space routinely, on a moment's notice if necessary, and at a small fraction of today's cost. This will be done within the framework of a useful total space program of science, exploration, and applications at approximately the present overall level of the space budget."[83] At the time of Nixon's statement, certain details of the architecture of the shuttle had only recently been decided: having prepared several possible models of the vehicle, Fletcher mistakenly brought with him a model employing liquid fuel rather than the preferred solid-fueled booster rockets, but Nixon was so enthusiastic about it anyway that he neglected to return it afterward.[84]

Having largely won a battle with NASA over the design of the shuttle, the Air Force reacted to Nixon's decision authorizing it with a "neutral recommendation" for its construction, undermining the agency's efforts. While some in the Air Force, like former Apollo manager general Sam Phillips, were enthusiastic, Seamans's support had never risen above a "qualified endorsement," contingent on the shuttle meeting "necessary performance requirements" and being "more economical to use than existing systems." Otherwise, the Air Force would employ expendable launch vehicles as it had always done, leaving NASA to fail on its own. Importantly, Air Force engineers still recommended against the inexpensive solid rocket booster (SRB) technology that NASA favored for the shuttle, but that the Air Force feared was too unreliable for regular use on a piloted craft.[85]

Nixon's meandering January 5 statement matched the ambivalence not only of the Air Force but of the American people. Indeed, in lieu of actual public support for the space shuttle, opinion research after Nixon's announcement demonstrated that, like the president, a "plurality" of Americans held favorable views of spaceflight generally, but not of any particular spaceflight program. A September 1972 study commissioned by NASA from the Opinion Research Corporation of Princeton, New Jersey, found that among survey respondents (including a large percentage holding favorable impressions of space exploration), support for specific NASA programs like space stations, deep space probes, and space shuttles remained below 15 percent, while a return to the Moon excited a whopping 4 percent of respondents.[86] Oddly, whatever enthusiasm for spaceflight that existed largely manifested, not in a concern about military assets or low-cost access to orbit, but in the belief held by a large number of respondents that the shuttle would solve the problems of pollution and environmental degradation, which were already significant public concerns. Although the source of this belief is unclear, Nixon disingenuously encouraged it, promising the shuttle as the answer to "pollution," its creation motivated by the "imperatives of universal brotherhood and global ecology." The new spacecraft, Nixon promised, would somehow help Americans "think and act as guardians of one tiny blue and green island in the trackless oceans of the Universe."[87] The shuttle's principal purpose, to launch spy satellites, escaped mention—instead, NASA offered promises calculated to appeal to California voters in particular, to justify the development of a largely unnecessary new military space infrastructure.

No Escape

The TAOS design produced by the winning bidder, North American Rockwell, featured conventional wings and a payload bay large enough to accommodate military satellites (figure 3.1). Rather than featuring the straight shape favored by Max Faget, the shuttle's wings were triangular, in a delta configuration that improved cross-range at the expense of low-speed handling and landing performance.[88] Instead of a metallic heat shield, a mosaic of tens of thousands of ceramic tiles would cover the exterior of the craft. SRBs would boost the shuttle into orbit, their thrust supplemented by the shuttle orbiters' engines, fed from a giant external fuel and oxidizer tank. Only the orbiter would have fly-back reusability: the External Tank (ET) would reenter the atmosphere and burn up, while the depleted solid rocket motors would fall into the ocean beneath parachutes, be recovered by ship, and be refurbished for later flights.

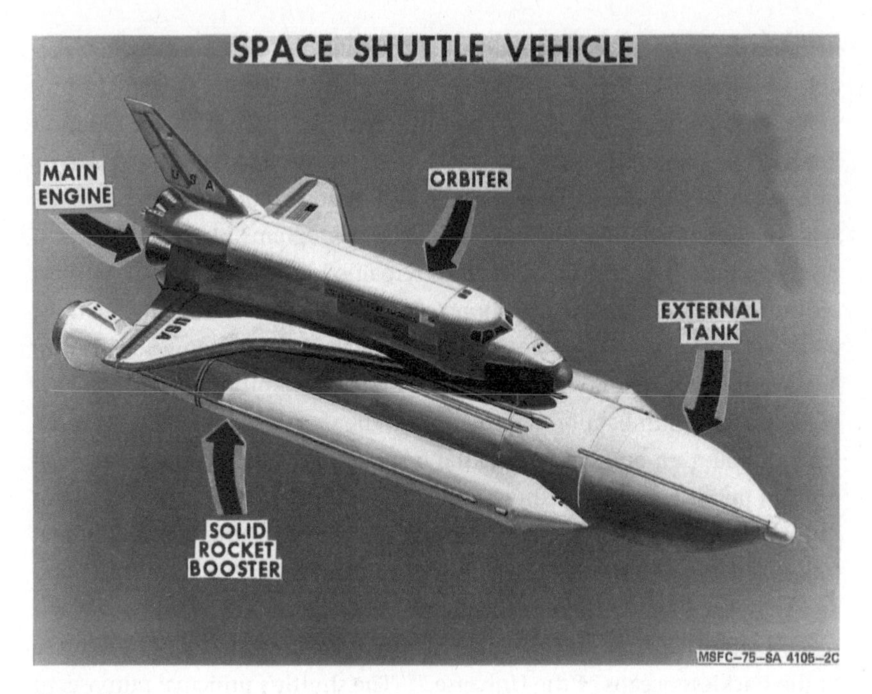

Figure 3.1
A compromise enabling partial reusability, the TAOS concept would boost the orbiter into space using reusable booster rockets and an expendable tank feeding the orbiter's own engines. This drawing by NASA's Marshall Space Flight Center (MSFC) depicts the space shuttle's 1975 configuration, including twin, parallel SRBs, the orbiter's three Space Shuttle Main Engines (SSMEs), and the twin Orbital Maneuvering System (OMS) engines used for final insertion of the craft in orbit, as well as its deorbiting and reentry. NASA photo.

Faget recognized immediately that this shuttle was a compromise (and a rather unpleasant one) and spoke of it in those terms throughout its development. "The basic design was established after several years of extensive study in which cost factors of the various desirable features and performance capabilities were examined," he noted in 1980, adding: "Although complete reusability was an initial design goal, this approach was abandoned because such a vehicle would have been larger, more complex, and much more costly to develop than the adopted approach."[89] While reusability was subject to compromise, the shuttle's size was not: whatever NASA built needed a voluminous payload compartment to accommodate Air Force spy satellites, as Air Force participation was essential to the program. Faget could not speak publicly about the shuttle's military role but alluded to it at industry conferences.

Photographic reconnaissance satellites contained reflecting telescopes that were massive in size but consisted mostly of empty space, requiring a shuttle configuration that could handle such payloads. Stated Faget, "This configuration is based on a compromise satisfying a large number of requirements. The cargo bay size was derived from a survey of potential traffic which clearly indicated that space cargo would consist in general of low-density objects."[90]

To conserve funds for the construction of such a large orbiter, NASA abandoned features intended to make its operation safer and more economical. Air-breathing engines, intended to increase the orbiters' cross-range and enable ferrying flight between landing and launch sites, were deleted from the design due to the cost and weight that they would add. The omission of these systems was critical for the shuttle program's survival, conserving funding for critical elements like the reusable engines and TPS. Once the shuttle flew, NASA hoped, upgrades would slowly transform it into a more sophisticated craft, or possibly even the fully reusable vehicle that NASA preferred.

The likely failure modes for the new vehicle—both economic and technical—were obvious. The existing market for launch services was so small that the economics of the shuttle seemed likely to falter after only a few years' operation unless declining launch costs suddenly created unexpected demand. Before economics doomed the spacecraft, though, the craft's structural compromises would likely catch up with it. Unable to manufacture SRBs that were large enough, North American Rockwell's booster subcontractor, Thiokol, would manufacture them in segments pinned together. Segmented SRBs had never been used on piloted space vehicles and had a reputation for structural weakness, vibration, uneven burning, and general unreliability: it was impossible to throttle them or shut them down in flight, and while the Air Force had attempted to qualify them for human (man-rated) flight, that program had shut down before certification. The side-by-side configuration of the orbiter and boosters, intended to enable the shuttle to use its own engines for liftoff, also exposed the crews to impact from ice and insulation ejected by the ET on liftoff, likely peppering the orbiter with debris and punching holes in its TPS. The shuttle had no reliable redundancies for breaches in the SRBs or punctures of the craft's TPS; either could destroy the orbiters.

Despite its appearance of robustness, the shuttle lacked any real escape systems, a problem that NASA engineers, consumed by cost and weight limitations, repeatedly considered but had never been able to address adequately. Theoretically, the two members of a flight crew seated on the flight deck could egress using ejection seats during the first few minutes of launch or the last few minutes of landing, but at high speeds, bailout from the shuttle would be fatal, exposing the exiting crew members to hypersonic wind blasts, frictional heating, burning propellant gases, and a vacuum that would asphyxiate the crew members and render parachutes ineffective. Once in orbit, ejection seats and parachutes would be useless, and a shuttle orbiter with a damaged TPS would not survive long enough in the atmosphere for ejection seats to be effective on landing. Shuttle astronaut Robert Crippen found ejection escape scenarios "unrealistic" and speculated that the two ejection seats added to the first fully operational shuttle orbiter, Columbia, were merely an affectation for former fighter pilots accustomed to their presence in the cockpit, or a gimmick intended to assuage fears that the shuttle offered little hope of survival in the event of trouble.[91]

One plan, to mount an Apollo-style capsule in the cargo bay in the event that the shuttle's TPS was damaged, NASA rejected for the additional weight that it would add to the vehicle. In the absence of such a pod, the only hope for rescue of a shuttle crew in orbit was to launch another shuttle to retrieve them. This was easier said than done: shuttle orbiters possessed airlocks but could not dock with each other. Given the location and configurations of its airlock (inside the cargo bay), the crew of a stricken shuttle would need to spacewalk (extravehicular activity, or EVA) to the rescue vehicle. NASA though, did not plan to provide expensive, delicate EVA suits for every crew member. Two fully trained pilots would crew each mission; the remaining five crew members could be any number of nonpilot career astronauts ("Mission Specialists") and guests ("Payload Specialists"), the latter having varying levels of training. With only a small number of EVA suits and fully trained crew members onboard, NASA originally expected the majority of the shuttle crew members to wait out any emergency while zipped into a three-foot-wide Personal Rescue Enclosure (PRE), a technology that, like the shuttle, gave the appearance of efficacy but soon became emblematic of the failed efforts to build safety mechanisms into the craft.

Developed first by Ling-Temco-Vought in 1973 and later refined by spacesuit-maker ILC Dover under NASA contract, the PRE was an inflatable, zippered white beach ball approximately three feet in diameter, in which space-suited crew members would encapsulate their unsuited colleagues (figure 3.2). Once sealed inside, astronauts would breathe oxygen through an umbilical cord attached to the shuttle orbiter's environmental control system, or from a portable oxygen system providing sixty minutes of breathable

Figure 3.2
Emblematic of NASA's sluggish embrace of women astronauts, the six women that NASA selected in 1978 as Mission Specialists appear in this 1979 publicity photo with the PRE, a proposed emergence egress device into which the women would pack themselves if the need arose for them to leave a stricken shuttle while in orbit. From left, Margaret "Rhea" Sedona, Kathryn Sullivan, Judith Resnik, Sally Ride, Anna Fisher, and Shannon Lucid. NASA photo.

air.[92] Before time ran out, pressure-suited crew members would restore the cabin pressure or transfer the neatly packaged crew members to a safe location via an attached carry strap. NASA publicized this morbid technology heavily in the years up to the first shuttle flight (often in connection with women astronauts),[93] but while the technology of the PRE appeared feasible, its likely mode of use was improbable. Encapsulated crew members simply had no place to go, as NASA was unable to earmark the funds necessary to ensure that a second shuttle remained always at the ready to rescue them.[94] NASA's promotion of the PRE ceased, it declined to order production units, and the "rescue ball" eventually found use only as a screening tool in astronaut selection to weed out claustrophobic applicants.[95] Critically, the decision to abandon the PRE did not make astronauts any safer; instead, NASA traded an imperfect on-orbit rescue technology for none at all.

Conclusion

Created in the belief that that the US must both continue its exploration of space and broaden popular participation in that effort, the space shuttle seemed to fit comfortably into the "winged gospel" of aviation's transformative power. That NASA would follow Project Apollo with some kind of space program was a given among the nation's political leadership in 1972, but arguing most significantly for the shuttle's development was the fact that it was far cheaper than most alternative proposals, especially an Apollo-style mission to Mars.[96] Of the panoply of space vehicles and projects NASA considered undertaking after Project Apollo—including space stations, nuclear orbital ferries, Venus fly-bys, and Mars landings—the shuttle offered the lowest costs and the most immediate utility, although its value hinged upon unrealistic economic assumptions about a future in which spaceflight would be almost as routine as airline travel.

Spacemindedness required an inexpensive vehicle, and NASA's commitment to a reusable, winged craft left it little room to entertain plausible alternatives: a new, pernicious winged gospel that undermined the shuttle's reusability and safety. The TAOS vehicle that NASA eventually decided upon constituted an uneasy combination of the most necessary design features, twisted to suit national security needs, with the promise that refinements and improvements would follow once funds became available. With its manifold faults, the most that could be said for the shuttle's design, and indeed, the entire shuttle program, was that it was the cheapest option that gave the *appearance* of being able to fulfill the role assigned to it. "As the other elements of a big human program faded into a vague and distant future," Michael Neufeld concluded, "the Shuttle became for a decade an end in itself—not so much a space policy as an excuse not to have one."[97] After NASA and the Air Force settled upon the design for a partially reusable craft, expected improvements never arrived, and the system quickly became both technically risky and too expensive to operate.

4 Tickling the Dragon, 1974–1986

The Allied effort to build the atomic bomb during World War II exposed thousands of scientists, engineers, and workers of all kinds to unprecedented risks, many of which they lacked the knowledge and tools to control adequately. The most essential component of the bomb—fissile uranium or plutonium whose atomic nuclei would split when struck by neutrons—was hazardous to mine or fabricate and fashion into a bomb core, and an atomic bomb's yield was at best an educated guess.[1] Edward Teller speculated that the weapon might be powerful enough to set fire to the atmosphere and destroy humanity, but later calculations put that idea to rest.[2] Most scientists expected less from the "gadget," but measuring the yield of atomic weapons without detonating them required a kind of experiment that was dangerous enough to kill those who attempted it: assembling a near-critical mass of fissile material just long enough to begin a fission chain reaction, in which the splitting of atomic nuclei released neutron radiation that induced nearby atoms to split as well. Atomic bomb cores could not sustain controlled chain reactions: taking a mass of plutonium designed to detonate completely in a fraction of a second and teasing it just enough to spew a few neutrons was the physical equivalent of what the Manhattan Project scientist Richard Feynman called "tickling the tail of a sleeping dragon."[3]

During the 1940s, such experiments were conducted under circumstances that, even then, were regarded as shockingly unsafe: scientists wore no protective clothing and coaxed bomb cores to fission by manipulating the devices by hand.[4] The scientists knew that each time they performed one of these tests, they risked a criticality accident that would spew enough neutron radiation to kill those nearby.[5] On two occasions, they did. On August 21, 1945, three days after V-J Day, the physicist Harry Daghlian, Jr., received a fatal dose of radiation while performing a criticality test on the plutonium core of the third atomic bomb. The following year, the Canadian physicist Louis Slotin died from neutron radiation exposure while conducting a similar experiment on the same core. Attempting to keep two hollow hemispheres of a beryllium neutron reflector apart with a screwdriver, the tool slipped, allowing the hemispheres to close around the central plutonium sphere, reflecting stray neutrons back into it and accelerating the core's fission reaction. The burst of radiation emitted from the core ionized the surrounding oxygen and turned the air blue. To save his colleagues, Slotin pried the neutron reflectors apart with his bare hands, further dooming him to a painful death.[6]

While the dangers posed by radiation exposure were not as well understood in 1946 as they were decades later, scientists recognized the potential lethality of the experiments and made some efforts (however insufficient) to protect themselves. Tickling the dragon was never regarded as a safe activity, but its perceived necessity in the last year of World War II and the early years of the Cold War outweighed concerns about its safety, and Manhattan Project scientists made relatively little effort to obviate the need for such experiments or make them safer. Indeed, they could do neither: the atomic bomb's instability was the reason for its existence, and it could never be anything but a menace. The danger did not lie in the tickling, but in the dragon itself.

During the early years of space shuttle operations from 1981 through 1985, the National Aeronautics and Space Administration (NASA) similarly undertook a series of actions that, to later scholars, appeared to grease the skids of the space shuttle Challenger's doomed flight in January 1986. These evaluations, though, are probably unfair: the space shuttle was, throughout its history, always a political kluge and a work-in-progress: an experimental craft with known vulnerabilities whose complex problems evaded simple solutions. Compromise and thrift had driven its configuration, and its principal benefit lay in its authorization in an election year. The shuttle's construction, President Richard Nixon hoped, would ensure American preeminence in space at the lowest possible cost, enhance America's technological and economic strength (or at least appear to do so), and, most important, guarantee aerospace employment for years to come. Although not publicly known, the shuttle also marked a blending of NASA's and the national security community's parallel but distinct space programs, enlisting civilian resources in spy satellite deployment and other national security missions.

For a while, the shuttle appeared likely to deliver on its promises, but only if one did not look too closely at it. Indeed, the shuttle's construction in the 1970s and its early missions of the 1980s produced a broad recognition of the vehicle's flaws by both NASA and its contractors. The apparent ease with which NASA flew the shuttle in 1981 and then accepted it as an operational transportation infrastructure in 1982 owed less to confidence in the craft than to sheer fatalism: spaceflight had always been dangerous, and modest changes were unlikely to affect the likelihood of a catastrophic mishap. NASA, throughout this period, tickled the dragon—stripping spacesuits from shuttle crews, flying civilians in public relations campaigns—but none of these actions made much difference in either augmenting or diminishing the shuttle's dangers. Whether tickled or not, the dragon would roar.

Assembly, Rollout, and Testing

As with Project Apollo of the 1960s, the construction of NASA's fleet of shuttles fell to the prime contractor, North American Aviation (responsible for much of the fuselage and the integration of various subsystems), and to a network of subcontractors stretching across the US. Construction of the first orbiter began at Rockwell's North American plant in the Los Angeles suburb of Downey in 1974. Companies in congressional districts across the country manufactured the wings, rudder, and other portions of the orbiters to ensure the shuttle's support by a diverse array of lawmakers, and then shipped the parts to Rockwell's plant in Palmdale, California, for final assembly. Once it had the external appearance of a functioning spacecraft, the first orbiter would arrive at NASA's nearby Dryden Flight Research Center to verify its airworthiness by releasing it from a carrier aircraft for gliding tests. Even as late as mid-1976, though, significant technical problems with the design remained, particularly regarding the engines and thermal protection system (TPS). So did questions about the shuttle's practical uses and the extent to which other nations would participate in the Space Transportation System (STS) program, one of the earliest justifications that the Nixon administration had used to argue for the shuttle's existence. Rather than afterthoughts, these questions had bearing on the architecture of the STS and required immediate answers.

To Nixon, who tended to view the shuttle in political terms, the shuttle's role as an instrument of international diplomacy was one of the few areas in which he assumed anything approaching a personal interest. Unlike Apollo's cramped quarters, the shuttle offered opportunities to fly foreign astronauts from allied nations in relative safety and comfort and confer public relations benefits on NASA that were potentially even greater than those of Apollo, embarrassing the Soviet Union in the process. John Logsdon notes that NASA, when informed of Nixon's interest in the shuttle as a tool of diplomacy, likely overestimated it, assuming that Nixon supported European participation in manufacturing the vehicle, not just in crewing it. (NASA had believed, for example, that Nixon would permit shuttle subassemblies to be contracted out to foreign companies, a gross misreading of Nixon's political priorities.)[7] Most appealing to NASA's international partners, though, was the chance to reap the profits for building other elements of the STS, especially an autonomous craft that would move satellites from the shuttle's orbit to higher ones: a "space tug."[8] Another option was the concept of the "Sortie Can," a pressurized laboratory mounted in the shuttle's cargo bay to be used by scientists.

As shuttle development proceeded, concerns about cost and American control of cutting-edge space technology began to complicate the shuttle's role as a tool of international diplomacy. The space tug would be particularly critical for the launch of national security payloads, and Nixon's advisors warned that the funds that European nations contributed to this endeavor would purchase access to the shuttle's complex technical know-how for a relatively small investment. At one time enthusiastic about international participation, Nixon, by 1971, had already confessed to aides that he sought only the "'symbolism'" of international participation and backed away from intensive European participation in the shuttle program.[9] By 1972, even NASA was, rather than pleased by the support that its European partners gave to the program, concerned that it was "stuck" with them, unable to separate itself from partnerships not in NASA's economic interest.[10] Within the next few years, most European participation in the STS faded, with plans for a flying laboratory (renamed Spacelab) maintained as the only significant European contribution to the program. A combination of pressurized modules and external equipment panels, the laboratory increased the internal volume of the shuttle's crew compartment while in orbit but could not operate outside the confines of the shuttle's cargo bay (figure 4.1). Completed after the orbiters, it remained a source of controversy throughout the program.

Spacelab came to epitomize the challenges involved in turning NASA's hazy notions of the shuttle's function into a working spacecraft enjoying broad support. Behind the negotiation and collaboration required to solve Spacelab's diplomatic challenges, NASA personnel expressed misgivings about

Figure 4.1
This January 1981 artist's conception depicts the European Space Agency's Spacelab modules, including a pressurized laboratory, in the cargo bay of a space shuttle orbiter. NASA photo.

accounting irregularities and optimistic assumptions supporting its construction that would soon engulf the entire shuttle program. As with other prospective shuttle payloads, Mathematica analyses used to justify Spacelab's development projected twenty flights a year, a number at which many in the scientific community scoffed. Noted Gregg Easterbrook in a 1980 exposé in the *Washington Monthly*:

> Will Spacelab be used 20 times a year? "I can't imagine what for," says Albert Cameron, a Harvard physicist who is chairman of the Space Science Board of the National Academy of Science. "The duration of the flight is so short," he explained, pointing out that it will ordinarily last only four to seven days, "there's way too little time to carry out any meaningful experiments." Even [Spacelab director Doug] Lord acknowledges that the applications of Spacelab are limited: "It's really an interim step, to demonstrate to the world that a permanent space station is a worthwhile idea." Meanwhile, will the interim step fly 20 times a year—a total of 200 times? "Not a chance," says an informed NASA source.[11]

Spacelab was, simply, not sufficiently useful to justify its cost as a symbolic diplomatic tool. Even worse, NASA would pay for all but one of these flights, depriving the shuttle of revenue that other payloads might generate.[12]

As with the entire shuttle program, justifications for Spacelab's development made sense only after ham-fisted accounting tricks obscured outlays and program goals. When cost estimates threatened to undermine Spacelab, one August 1976 report to NASA administrator James Fletcher noted that NASA managers had already revised cost estimates downward based on less accurate experimental goals, confusing even program managers as to the vehicle's actual cost and utility:

> One of the problems that seems to afflict Spacelab is that the resources initially estimated for the payloads and related integration activities that have been defined were considered unacceptably high in many cases leading to assignment of lower funding levels with obscure program content. This process keeps the program definition in flux with loss in our ability to specifically explain and justify what we are doing and its worth.[13]

Massaging the figures to satisfy the Office of Management and Budget (OMB) or Congress was one thing, Fletcher feared; NASA would face real trouble if it could no longer keep track of its own accounting tricks.

Even more threatening to Spacelab than the discovery of its manipulated accounting was the fact that under even optimum circumstances, the science that Spacelab could perform (life sciences research and materials processing) was of little value to NASA and its international partners; reducing expectations even further did nothing to improve the rationale for the lab's existence. Indeed, one dog-eared page of an appendix titled "Status of Spacelab Payloads" noted: "ONLY LIFE SCIENCE AND SPACE PROCESSING REALLY NEED SPACELAB AND THEY ARE OUR WEAKEST DISCIPLINES"; an anonymous annotation (possibly by Fletcher) marked the statement with an arrow and the words "Key Page." For a system designed to lower the cost of space travel, growing costs and questionable utility made a potentially lethal combination: "We have other concerns," the August report noted, including "the complexity of the interfaces with possibility of funding duplication or exposure, and the estimated size of integration effort and support services (data processing, etc.) for what was basically promoted as a cost saving system."[14] Concerns about the shuttle's economics multiplied through 1976, along with growing efforts to work around the shuttle's known physical vulnerabilities. With congressional funding already approved and construction underway, NASA siloed questions about the shuttle's utility, largely ignoring them in the hopes that the shuttle's success would obviate most of the concerns.

Debates on the shuttle's purpose grew alongside controversies about what to call the vehicle, with both raging at the time of the shuttle's official roll-out. The naming issue, which NASA believed was a simple matter, eventually became a question for the highest levels of the executive branch. Intended for completion on Constitution Day, September 17, 1976, the first space shuttle orbiter would undergo a series of piloted drop tests from a carrier aircraft before returning to Palmdale for completion as a functional space vehicle. NASA had intended to call the first shuttle orbiter Constitution,[15] but it found unexpected opposition among one group that it had been slowly courting since the shuttle's announcement: fans of the short-lived 1960s NBC science fiction series *Star Trek*. When the fans of the canceled show first gathered in 1972, a solicitation to NASA for assistance resulted in the agency sending a trailer of samples and memorabilia to the event, and the two communities of space enthusiasts remained on good terms.[16] Science fiction enthusiasts were not acritical supporters of NASA, however; many advocated more extensive explorations than those contemplated by the agency, while others insisted that the public play a more direct role in the shuttle program.

Before Constitution's completion, a letter-writing campaign by hundreds of thousands of *Star Trek* enthusiasts pressured NASA to name the new craft Enterprise instead of Constitution, honoring the name of the TV series's fictional starship. Among the name's supporters was the science fiction author Isaac Asimov, who wrote to Fletcher on August 5, 1976, reminding him of the historical legacy of "Enterprise" as the name of a series of famous warships (including a storied Yorktown-class World War II aircraft carrier and a nuclear-powered aircraft carrier then in commission) and the fact that the Navy, when christening its first nuclear submarine, had poached the name of a fictional craft, Nautilus, from the work of the author Jules Verne.[17] President Gerald Ford, a former naval officer with a personal connection to the World War II carrier,[18] was enthusiastic about the proposed name change, and the fact that "enterprise" was also associated with capitalism made the moniker palatable to his administration despite its pedestrian association with a cult television show.[19] Attempting to turn the public relations flap into a victory, NASA assembled the cast of the original series and its creator, Gene Roddenberry, for the first public appearance of the shuttle.

The rollout of Enterprise in 1976 marked the closest thing to a public relations victory that the shuttle experienced during its development, but the event likely misled the public about the capabilities and readiness of the new craft. Identical to the eventual space vehicles in size and aluminum construction (though somewhat heavier), Enterprise had no main engines, maneuvering thrusters, or TPS (foam and fiberglass covered the fuselage and wings instead). Instead of a full cockpit, the craft offered pilots only rudimentary controls, which were intended to allow them to test the vehicle's performance upon landing. Instrument panels lacked all but the most basic indicators, and power and life support equipment was rustic, with pilots wearing jet aircraft–style oxygen masks during flight. Even stripped down, though, Enterprise reasonably approximated a real space shuttle returning to Earth with its cargo bay emptied and fuel depleted.

With no human space vehicle having ever landed on a runway after its flight, NASA initiated an elaborate test program to verify the half-finished orbiter's airworthiness, ferrying it to medium altitudes atop a converted Boeing 747 passenger jet to test its aerodynamic performance, and eventually releasing it, unpowered, to glide to a landing at Dryden. The shuttle's Approach and Landing Tests (ALT) occurred without incident through 1977, but widely circulated NASA photographs of the captured flight and drop tests confused popular audiences about the shuttle's flight mode.[20] Images suggested that the shuttle orbiter would launch atop its winged carrier, just as Max Faget had proposed with the DC-3 a decade earlier (figure 4.2). This myth was fueled by the dramatic opening scene of the 1979 James Bond film *Moonraker*, in which a space shuttle is hijacked and stolen in midair, blasting off the back of its carrier jet with its own rocket engines.[21] Shuttle orbiters, however, could never fly into space on the back of a commercial jetliner: the fuel for the shuttle's main engines would be carried externally during launch, and once separated from the External Tank (ET), the shuttle's engines were ballast. Yet NASA so often published depictions of Enterprise atop its carrier jet[22] and so frequently touted the shuttle's jetlike reliability and "airliner cockpit"[23] (resembling that of the then-popular DC-10 wide-body passenger jet)[24] that public relations materials increased rather than abated confusion as to how, exactly, the shuttle flew. Combined with fanciful popular depictions of space shuttles in film, these references established impossible popular expectations for the shuttle, including the frequency of launches and cost of its projected missions.

Figure 4.2
The space shuttle Enterprise (top) separates from its Boeing 747 carrier aircraft while undergoing aerodynamic flight testing over Dryden Flight Research Center in California in 1977. NASA photo.

As the ALT flights progressed, difficulties completing the most sophisticated elements of the shuttle's design kept pushing the date of the First Manned Orbital Flight (FMOF), which NASA intended to coincide with the tenth anniversary of Apollo 11's voyage to the Moon in July 1969. The appearance of completion demonstrated by Enterprise's rollout belied the true state of the shuttle program: external evidence of progress masking continued confusion and experimentation. The beginning of President Jimmy Carter's administration in 1977 forced another reevaluation of the shuttle program, but by then, concerns had shifted from whether the space shuttle would be ready to fly in 1979 to whether it should exist at all. Once again, executive branch voices outside NASA noted the craft's experimental nature and suggested that the shuttle fleet shrink, pending its replacement by a more robust craft, or at least a redesign of its more problematic systems.

NASA had originally proposed a fleet of five orbiters and two launch sites, in Florida and California, from which to embark them. At the launch rate projected by NASA, however, two orbiters and the one existing launch site at NASA's Kennedy Space Center (KSC) were sufficient to meet the demand for satellite launch services, suggesting that NASA's projected launch rate was purposefully inflated. The shuttle would not produce the desired economic efficiencies because the five orbiters would not have enough launch business to sustain their use. To shuttle critics in Congress and the Government Accounting Office (GAO) in 1977, the construction of additional orbiters appeared merely to be a way to fund long-delayed system upgrades by replacing older orbiters. New orbiters would receive the improvements that earlier vehicles lacked, and thus the shuttle was a phased development program, despite protestations otherwise.

As the GAO concluded:

> It is also questionable whether the present configuration of the STS is adequate to serve the needs of the nation during the 1980's[sic]. The STS is essentially a research and development program and will not meet NASA's goal of fully reusable launch vehicle. This, coupled with the limited maneuverability of the space shuttle, may result in the STS being upgraded, modified, or partially replaced. NASA is studying ways to upgrade the STS and is examining the feasibility of a fully reusable single-stage-to-orbit shuttle. A heavy lift launch vehicle is also being evaluated because the shuttle's capacity is not adequate for all expected applications.[25]

The GAO concluded that instead of rushing to build the Thrust-Assisted Orbiter Shuttle (TAOS) orbiter in quantity, NASA should instead limit production until the STS demonstrated its effectiveness. NASA was extremely hostile to such suggestions, for several reasons. Not only might the additional funds to replace the STS not be forthcoming, but the cost savings provided by the space shuttle system could manifest only with a fleet large enough to service the predicted growth in space launch traffic. As a means of meeting only NASA's existing launch needs, the two-shuttle system's high operating costs would demonstrate that development of the shuttle system had been a mistake, leading to its cancellation. Again, an incremental approach that would have allowed a reengineering of the shuttle prior to the loss of Challenger disappeared due to the threat it posed to the whole program.[26]

Fletcher, no longer NASA administrator, could not resist remarking on these issues in a 1977 letter to NASA deputy administrator Alan Lovelace at the time of the ALT program. Comparing shuttle development (which he had recently supervised) to previous NASA and Air Force programs, Fletcher scoffed that while the shuttle program sought to maintain a reliability in excess of 99 percent (exceeding that of previous, less complicated launch vehicles), NASA had not even maintained its former cadre of experienced engineers and managers familiar with the peculiar problems of rockets. Rather, it had relied upon a stable of airplane engineers familiar principally with a field in which the "design problems are entirely different." Fletcher offered no solution short of reconstituting 1960s design teams, which even he recognized as impractical. Instead, Fletcher mainly hoped to absolve himself of his lingering guilt about advocating for the shuttle and approving its risky design:

> When you put all this long-winded background together, what does it add up to? I'm afraid I don't have a good answer. I simply had to unload it on you partly to clear my conscience and partly to help make you especially cautious as you approach the FMOF date. I have suggested a quiet review which you did make, but only Walt Williams in that review group had launch vehicle background. Eberhardt [Rechtin], Kurt Debus, Ernst Stuhlinger and perhaps some TRW Systems types might be helpful. I'm afraid Lockheed or Boeing wouldn't be much help since Polaris and Minuteman are so much simpler in every respect than the Shuttle.
>
> Anyway, the problem is yours and [NASA administrator Robert A. Frosch's] now. I intend to say no more about it.[27]

Fletcher, who had stood beside Nixon only five years earlier celebrating the shuttle's birth, now washed his hands of it, concerned about flaws that other NASA administrators would need to confront.

While engineers at Rockwell and NASA continued to make incremental improvements to a variety of systems from electronics (the orbiters' onboard computers were notoriously bug-prone)[28] to the construction of the shuttle's wings, others wrestled with a problem so severe that it threatened the continuance of the entire program: the implementation of the shuttle's TPS. With NASA having fixed upon aluminum construction and a nonmetallic shield in 1972 to reduce the vehicle's cost and weight, engineers at Rockwell set upon developing a reusable ceramic material that it could plaster over the shuttle's airframe to protect it from the frictional heating of reentry, likely to exceed several thousand degrees Fahrenheit. This was a controversial technology: Lockheed had rejected it for its own hypersonic research aircraft (the Lockheed L-301) concerned about its excessive maintenance requirements and complexity.[29] The material would have to be durable and reliable, lasting

100 missions without the need for replacement and protecting the orbiters' aluminum fuselage for 500 flights.[30] The shuttle, like any metal airframe, would bend in flight under aerodynamic loads, making the application of any nonflexible covering impossible. For designers and technicians, the practical effect of these instructions was to require them to figure out a way to cover the shuttle's curved, flexible metallic fuselage and wings with sand.

Silica tiles consisting mostly of air could hold their shape, withstand heat, and extract little mass penalty, but they could not flex without breaking. To solve this problem, engineers covered the skin of the shuttle orbiters with blankets of flexible, fire-resistant Nomex cloth, and then divided the surface area into nearly 30,000 individual squares roughly the size of a child's peanut-butter-and-jelly sandwich.[31] Over each, technicians glued a single specific silica tile, cut to size, numbered, and fastened exactly on the designated spot using a silicone adhesive (room-temperature-vulcanizing [RTV] silicone). Sizing and placement were critical: gaps in the TPS would not merely allow heat to enter the orbiter and melt its aluminum superstructure; they also would disrupt airflow around the craft, potentially causing even greater damage. The spacing of even well-placed tiles would still create small gaps, so in certain critical areas of the spacecraft, workers stuffed fiberglass between the tiles like grout to smooth airflow and prevent heat intrusion. (These "gap fillers" were themselves prone to falling off and constituted an additional difficulty.) While more heat-resistant (but brittle) carbon panels protected certain areas, and flexible thermal blankets covered cooler spots, the silica tiles that covered the majority of the orbiters' surface posed the greatest manufacturing challenges (figure 4.3).

The choice of a ceramic TPS turned the construction of the most modern aircraft in history into an exercise in premodern craft practice.[32] It required the installation of a mosaic of 30,000 glued tiles, expected to survive under ridiculous pressures and temperatures. The process of installing the tiles was laborious handiwork: each worker, even at optimum speed, could not manage to install more than one per day, and many tiles could not be affixed to the orbiter with sufficient strength to survive prelaunch checks, requiring time-consuming repairs. Later problems with the tiles, Hans Mark recounted, were twofold: not only was the RTV adhesive too weak to hold the tiles on the shuttle under the aerodynamic stresses of launch, but even when the adhesive held, the tiles themselves often broke in two.[33]

Figure 4.3
Astronauts aboard the International Space Station (ISS) took this image of the space shuttle orbiter Discovery to verify the integrity of its TPS. In addition to the reinforced-carbon-carbon (RCC) nose, the shuttle's ceramic tiles are visible, as is at least one protruding gap filler. The variation in color among the tiles reflects the maintenance performed between missions to replace excessively worn or damaged tiles with new ones. NASA photo.

By 1979, the TPS became the pacing item in determining the shuttle's launch schedule, and the work installing tiles continued even after NASA delivered the first spaceworthy orbiter (Columbia) to the KSC in Florida that year.[34] Even with the delays mounting, NASA promotional materials promised that "by the mid-1980s," a fleet of four orbiters would make "about one roundtrip per week," an optimistic estimate hinging on a near-flawless performance of the TPS.[35] Concern about the slow pace of tile installation, though, combined with ongoing problems with the shuttle's main engines, prompted the Carter administration to consider canceling the shuttle in 1979. Discussions that year between Carter, OMB, and NASA scoured the program for delays and design deficiencies, and while NASA offered a cautious insistence that the shuttle would fly by 1980 or 1981, debates about the STS revealed internal concerns not only about readiness, but about the shuttle's overall riskiness.

Apollo 8 astronaut Bill Anders, OMB wrote, had warned of the lack of unpiloted testing of the launch systems and crew vehicle:

> One of the three senior outside consultants appointed to review the Shuttle program on behalf of the President, former Astronaut William Anders, expressed the view that the Shuttle system had narrower-than-Apollo safety margins because of reduced hardware qualification testing and lack of unmanned flight testing for the Shuttle program. NASA program management believes that the Shuttle compares more favorably to the Apollo program when examined in detail.[36]

Grudgingly, OMB and NASA acknowledged that the shuttle orbiters lacked the structural safety margins of Apollo, but defended the shuttle by claiming that Apollo, like the shuttle, had numerous single points of failure and had often been entrusted with far more dangerous missions than the shuttles would perform. Apollo's lunar flights had been profoundly risky, so even with the shuttle's design deficiencies, its relatively modest Earth orbit missions would offer no more risk than Apollo lunar flights (which had achieved an 86 percent success rate). The ability of the shuttle to survive repeated flights was doubtful but irrelevant at this stage. The shuttle's FMOF required only that its components work satisfactorily once, which was a safety standard that NASA could easily meet with expendable components replaced with each flight—but a grudging admission of the STS's overall lack of reusability. The shuttle's FMOF would be risky, but no more so than an Apollo mission.[37]

Continued the OMB report, "While there is little difference in design safety factors between Apollo and Shuttle, Apollo did conduct more testing beyond design limits and did launch some early flights unmanned. However, the Apollo moon missions had single point failure vulnerability and fewer abort options than Shuttle. In the judgment of NASA program management, some Apollo flights were considerably more risky than Shuttle FMOF."[38] OMB's evaluation did not dismiss the shuttle's vulnerabilities so much as excuse them, by making false comparisons to Project Apollo. Unlike the earlier spacecraft, the shuttle lacked Apollo's safety redundancies, including a reliable solid-fueled escape rocket to blast the crew capsule free of a stricken launcher, a second spacecraft that could be used as a "lifeboat" (the Apollo Lunar Module), or a 100 percent launch success rate extending over a decade.

A glowing endorsement of the shuttle's safety this was not, but with the US lacking any other human spaceflight program, alternatives to continuing the shuttle program were few and far between. "The notion," noted one 1979 National Security Council memorandum, "that we are forced for short term economic reasons to abandon a major area of endeavor in which we have achieved world leadership at great cost is simply not credible."[39] The justification reeked of a sunk cost fallacy, but it proved remarkably persuasive. With significant funds already expended, the Carter administration saw no alternative but to accept multiyear delays and cost overruns and continue work on the vehicles.[40] Inheriting a Nixon-era space project that he opposed and an economic downturn, Carter reluctantly chose to continue the STS program, not seeing the results of his commitment during his single term in office.

Extensive delays caused by the shuttle's assembly problems helped kill one of the more enterprising ideas of the program: using an early mission to visit the derelict Skylab Orbital Workshop. In orbit since 1973, damaged in a launch mishap but functional,[41] Skylab had been visited successfully by three Apollo crews in multiweek stays in 1973 and 1974. At the conclusion of the last piloted mission, Skylab IV, the workshop retained enough supplies to support an additional visit. (NASA, years earlier, had abandoned plans for further Apollo flights, donating leftover Apollo launch vehicles and spacecraft to museums.) Still in orbit and available for use, the station's greatest vulnerability was neither its mechanical integrity nor its stock of supplies, but the orbit in which it found itself, which was low enough that Earth's atmosphere would likely slow it sufficiently to deorbit it. Skylab had no rocket engines aboard, only momentum wheels powered by electric motors that, when spun faster or slower, would induce rotation in the station.

NASA's early plans for the shuttle included a 1979 flight (which was designated as the second one on some projected flight rosters) to attach a rocket module to Skylab that would boost it to a higher orbit so that future shuttle crews could visit it; without it, the 170,000-pound cylinder would make an uncontrolled reentry and return to Earth. By the summer of 1979, though, Columbia was still two years away from completion, and greater than expected solar activity expanded Earth's atmosphere, further dragging Skylab downward. All NASA could do from the ground was rotate the craft to alter the atmospheric drag upon it and attempt to direct its eventual reentry toward a less inhabited part of the Earth: although much of the station would melt or burn, large metal components would survive reentry heating. Attempts to land Skylab in the Indian Ocean were unsuccessful, and on July 12, 1979, Skylab reentered over Western Australia, causing no injuries but leaving the sparsely populated area littered with debris.[42]

While visits to Skylab would have extended the shuttle's capabilities (and possibly created an international space station decades before shuttle visits to the Russian Mir), the loss of Skylab was, if anything, a benefit to NASA's effort to manage Columbia's potentially perilous testing regime. The reboost mission would no doubt have been a challenging one for an early crew: the flight would carry the shuttle's first remote manipulator arm (known as the "Canadarm") and would have required an extraordinary degree of precision in navigating and piloting the new craft. Instead, the revised second flight would have less ambitious goals.

First Flight

Before the space shuttle, all American human spacecraft flew in unpiloted launch tests prior to carrying human crews. Many of these flights had been quite extensive, sending a remotely piloted craft filled with instrumentation (or in Project Mercury, chimpanzees) on simulated missions to verify the safety of the launch, life support, and recovery systems. Although Columbia could be fitted with an automated control system to enable it to be remotely flown and landed, the expensive launch would squander an opportunity to fly a crew and, NASA engineers feared, provide an insufficient evaluation of the flight characteristics of the craft on landing.[43] (Like any piloted airplane, the shuttle had unique flying qualities that human pilots claimed only they could adequately assess.) Instead, NASA decided to launch the very first shuttle stack with a crew of two pilot-astronauts, who would test vehicle systems in flight and then land the orbiter, completing a full mission profile.

Although the shuttle's piloted first flight appears to have been perhaps the largest gamble in the history of American human spaceflight (equal to or exceeding Apollo 8's circumlunar mission on the first piloted Saturn V rocket launch), NASA's desire to place human pilots aboard the craft was described at the time as a corrective for the unreliability of its automated systems. According to one 1976 Johnson Space Center (JSC) safety analysis:

> Manning of the shuttle first vertical flight increases the probability of mission success and decreases the probability of vehicle loss. Man in the loop provides significant backup capability in evaluation, checkout, and operation of shuttle subsystems. Manual control during ascent or entry could prevent loss of vehicle in the event of a failure in the primary automatic systems. Reconfiguration of essential systems or overriding of automatic systems could be accomplished if conditions

are not consistent with premission planning. Mission completion could be accomplished in the event of a total loss of communications.[44]

Arguments like these had been present in the space program since its inception, typically employed by the Air Force and its pilots to discourage further automation of flight control systems and the retention of human piloting in new vehicles.[45] In NASA's estimation, the shuttle was more likely to succeed on its first flight with a crew than without it.

Commanding Columbia on its first flight was John Young, the veteran of four previous spaceflights and the last American astronaut on active duty who had walked on the Moon. Joining him in the right seat was Robert Crippen, a rookie who had transferred to NASA from an abortive military space program and had the misfortune of joining the astronaut corps in 1969, behind a backlog of astronauts selected in 1965, 1966, and 1967. Columbia lifted off on its first flight on April 12, 1981, in a mission designated as STS-1.[46] The launch attempt had been the mission's second; computer problems—a long-standing problem with the STS—had delayed a previous attempt. Immediately upon launch, the enormous ET peppered Columbia with debris. An orange layer of insulating foam covered the cold metal tank, but the foam usually collected atmospheric condensation on the launch pad, which would freeze and create an unstable surface that crumbled as the vehicle rose in flight. NASA had ordered the foam painted white to match the rest of the shuttle stack, and when white streaks appeared on the cockpit windows during ascent, engineers realized that the paint covering the tank had begun to shed. Shedding of paint, ice, and insulation had occurred on previous launch vehicles using cryogenic propellants (including those of Projects Mercury and Apollo), but these vehicles mounted the crew capsule safely above the debris path.

Other problems soon emerged, including with the TPS. The aft windows in the shuttle's crew compartment faced the internal cargo bay; after the crew opened the cargo bay doors, Crippen could see Columbia's rear fuselage and immediately noticed that dozens of tiles on the maneuvering pods were missing. Without communications with Earth for up to one-quarter of their time in space, the crew waited until they were within range of a ground station to share the troubling news.[47] Although Columbia carried two spacesuits for use in on-orbit emergencies, insufficient handholds on the smooth fuselage of the orbiter would make approaching many areas of damage virtually impossible.[48] Young and Crippen resigned themselves to their inability to repair or replace any tiles that might have fallen off the more critical spaces beneath the airframe. Recalled Crippen, "I know there was a lot of consternation going on the ground about it about . . . are the tiles really there . . . but there wasn't much that we could do about it if they were gone."[49]

NASA, though, preferred to know if a life-threatening emergency was present (if only to prevent it from reoccurring in the future) and had plans well before launch to order the crew to alter Columbia's orientation so that it would be viewable by a National Reconnaissance Office KH-11 Kennen spy satellite.[50] Although designed to inspect ground objects, its magnification and resolution would determine whether tiles had fallen from the orbiter's underside, even from miles away. Unlike the previous generation of American photographic reconnaissance satellites, the brand-new KH-11 employed an all-electronic imaging system the charge-coupled device (later employed in civilian digital cameras), that enabled the satellite to film live video using its telescopic cameras and relay it by radio directly to Earth ground stations. Real-time inspection verified that the critical portions of Columbia's TPS underneath the orbiter were intact, and that reentry and landing could proceed as scheduled two days later after thirty-six orbits of the Earth at an altitude of 133 miles.[51] Although Columbia, under a combination of automatic and manual control, did land successfully, heat had taken its toll on various subsystems, breaching the landing gear compartment and buckling the door (figure 4.4).

Encouraged by the performance of Columbia despite the in-flight problems, NASA continued its planned sequence of test flights, flying Columbia again in November 1981 with Joe Engle and Richard Truly, in March 1982 with Jack Lousma and Gordon Fullerton, and in June of that year with Ken Mattingly II and Henry Hartsfield, with the final missions extending over a week. Upon the completion of the last of these four flights, NASA declared the STS to be an "operational" rather than an "experimental" system, a classification that did not reflect any change in the shuttle's technical capabilities but would allow it to transition from short flights with skeleton crews to full missions embarking Mission and Payload Specialists. The shuttle's operational status was essential to NASA's ability to fulfill its commitment to the US Department of Defense: once routine, shuttle flights could contain classified military experiments and satellite payloads. Only an operational shuttle, furthermore, could fly commercial payloads, including privately owned communications satellites.[52]

This last goal of the shuttle—direct facilitation of private space commerce—was a new one for NASA and American space policy.[53] The launch

Figure 4.4
STS-1 pilot Robert Crippen descends from Columbia following its first flight; Commander John Young waits at the bottom of the stairway. The orbiter's thermal protection tiles, clearly visible in the photograph, were an early subject of concern for the mission because once in orbit, the astronauts observed that several had fallen off during the launch, and most were not visible to the crew, given the limited number of windows on the craft. NASA photo.

of Columbia in 1981 had coincided with the arrival of the administration of President Ronald Reagan and a decade marked by two paradoxical political trends: skepticism about the size and role of government—described by Reagan as "the problem" and not the "solution" to the nation's ills—and enthusiasm for massive government subsidies of private businesses, especially defense contractors.[54] A self-described conservative, Reagan turned out to be an ardent anticonservative when it came to military purchases and public funding of aerospace corporations; his enthusiasm for space travel tracked his lifelong interest in space warfare. Reagan was infatuated with science fiction books and films and enthralled by the possibility of future combat against extraterrestrials, even broaching the possibility for joint armed defense against space aliens with Soviet general secretary Mikhail Gorbachev.[55] These speculations alarmed Reagan's aides but did not manifest in actual programs; Reagan's White House subordinates made policy,[56] and their principal goal was to shift federal funding from nondefense programs to military ones,[57] with certain expenditures earmarked for major Republican donors.

Assisting private corporations through taxpayer subsidies—an anathema to Dwight Eisenhower in 1961—was the goal of Reagan's space policy and a departure from his supposed conservative beliefs, especially the need to free markets from the hand of government. Reagan's choice of NASA administrator reflected this view. Passing over longtime technical experts like Hans Mark, Reagan nominated James Beggs, an executive vice president of the defense contractor General Dynamics. Beggs had risen to the top of the list of candidates for NASA chief despite a lack of scientific or engineering credentials; like Reagan, though, he was fond of wearing cowboy hats.[58] In 1972, NASA had pitched the shuttle's revenue-generating features as one of many arguments for its approval; more friendly to space commercialization than his predecessors, Beggs moved its facilitation to the forefront of NASA's justification for the shuttle.[59] If federal agencies like NASA could not adapt to the new mandate to expand defense spending, they could at least utilize their appropriations to support the needs of American businesses or turn a profit themselves by using their infrastructure on moneymaking ventures.[60]

Encouraged by the shuttle's early flights and under pressure to show results, Beggs announced in 1982 that the shuttle was a functioning transportation infrastructure: NASA would offer inexpensive crew seats "on a reimbursable basis to all classes of . . . foreign and domestic commercial customers, international cooperative partners, the scientific and applications community and the Department of Defense."[61] Within a few years, Beggs hinted, the shuttle's corporate and military customers might even be joined "by U.S. citizens as passengers."[62] NASA would not only cultivate a commercial space industry, it would transform itself into a profit-making business by flying satellites, manufacturing facilities, and possibly even tourists into space.

These dreams hinged upon the availability of a cheap and reliable space vehicle, but the shuttle in 1982 was anything but. Not only did NASA severely underestimate the cost of each shuttle launch in its pricing models, but it gave steep discounts to military and intelligence users and faced competitive pressures from cheaper, unpiloted foreign launch vehicles like the European Space Agency (ESA)'s Ariane.[63] Even had it worked, Beggs's plan to rent the shuttle to business users did not stem from either Eisenhower's conservatism or John F. Kennedy's liberalism, but rather from pure crony capitalism: instead of exploring space, NASA would replace the embryonic free market for launch services with a single government provider that purchased expensive, unreliable rockets from key defense contractors selected by political appointees, and then priced their flights below cost for favored users, destroying the competitive pressures that might have improved the technology.[64]

The Accelerating Tempo

NASA's decision to declare the space shuttle operational in 1982 after only four test flights later attracted significant criticism, with many scholars writing that the shuttle never actually achieved true operational status.[65] Throughout the early 1980s, NASA's four flightworthy shuttle orbiters (Columbia, Discovery, Atlantis, and Challenger) underwent extensive modification, repair, and maintenance between flights, and subsequent orbiters included a number of changes based on lessons learned from Enterprise and Columbia. Declaring the shuttle operational, though, was essential to attracting customers for the vehicle, and thus NASA's decision to do so was as much political as technical, or even more so, as any perception of risk in the system poisoned any efforts to promote it to military, corporate, and international users.

NASA had no need for an experimental, two-person Earth-orbit spaceplane with no payload, and further flights of Columbia as configured for STS-1 through STS-4 would have served little purpose and likely would have been canceled by Congress. Ideally, NASA would have placed the entire STS program on hiatus after STS-4 and retooled the design after another five-to-seven-year hiatus. Absent a political mandate to do so, NASA carried on with the vehicle that it had. Calling the shuttle operational did nothing to ameliorate its design problems; NASA's failure, as Feynman later wrote, was to create false public expectations of the reliability of an inherently unreliable transportation system, while working aggressively behind the scenes to keep it safe. NASA knew full well about the dragon's ferocity, but it was unable to leave the tail alone or else risk falling out of the human spaceflight business entirely.

Critical design flaws, a paucity of realistic abort modes, and expensive maintenance problems with the space shuttle existed from the start of the program; engineers studied and debated them, but the complexity of the system offered no easy or cheap solutions. The shuttle's ET was literally a flying bomb of liquid hydrogen and oxygen, while the orbiter's tiled "skin" was so fragile that it could not withstand high-speed impacts by virtually anything and still maintain its structural or thermal integrity. The Space Shuttle Main Engines (SSMEs) were engineering marvels that operated under tremendous pressures and were subject to enormous wear and tear. Other sources of concern included the solid rocket boosters (SRBs), each of which consisted of cylinders of explosives poorly joined to make one large rocket that could fragment and destroy the rest of the shuttle. While many observers suspected that a catastrophic failure was possible, or even likely, astronauts and engineers did not agree on where it would happen first: the shuttle had too many failure modes to say for sure. Warnings about the shuttle's complex main engines, unreliable SRBs, and vulnerable TPS were commonplace in professional journals and the popular press through their development. So, too, were concerns that that shuttle's completion was accelerated with the expectation that a safer, more fully reusable booster technology would replace it, finally generating the system's supposed benefits.[66]

Following the loss of Challenger in 1986, commentators found ample evidence that NASA's risk tolerance had led it to make what appeared to be unwise choices in managing the program. Whether these decisions materially affected safety is unclear: given the shuttle's architecture, the catastrophic destruction of an orbiter on launch or landing was not a recoverable event for the crew, and NASA could do very little to make it one. The longer NASA flew the shuttle, the greater the certainty of loss; but beyond that, minor safety improvements mattered little. For astronauts, fatalism, rather than risk calculation, was the order of the day. "Every astronaut knew what the shuttle was," Mike Mullane (2006) later wrote, "a very dangerous experimental rocket flying without a crew escape system."[67]

By the time of the fourth shuttle flight, Columbia's vestigial ejection seats were still live, and the STS-5 crew weighed the merits of keeping them for their upcoming mission. Commander Vance Brand and Pilot Bob Overmyer insisted that ground crews pin the seats to prevent the ejection rockets from firing; Mission Specialists Joe Allen and Bill Lenoir (who would not be sitting in ejection seats) suggested that the seats remain in usable condition, and two survivors in the event of mishap were better than none. Brand, though, was adamant, claiming to be unwilling to survive if his crewmates could not. The shuttle program manager and astronaut Deke Slayton ultimately decided the question, insisting, "Everybody's going to have an ejection seat or nobody's going to have one."[68] The ejection seats were eventually disabled; Allen, in a later interview, acknowledged Brand's concerns, citing research from World War II suggesting that bomber pilots who had bailed out under similar circumstances had suffered a lifetime of emotional distress.[69] The astronauts' assent to removing safety redundancies from the space shuttle was not an expression of their confidence in the vehicle, though, but rather a lack of confidence in the safety systems themselves. They believed that the devices were unlikely to work when needed and would add complexity to already dangerous missions.

Another controversial NASA decision—to send shuttle crews aloft without spacesuits—smacked of recklessness but only marginally increased the astronauts' chances of injury in the event of in-flight mishaps. Since 1961, all of NASA's piloted vehicles had pressurized cabins, but astronauts had worn full pressure suits in case of emergency decompression, and later, to facilitate extravehicular activity (EVA) and lunar surface exploration. As vehicle reliability and internal volume increased, astronauts had been able to remove first their helmets and gloves, and then their entire pressure suits, swapping them for flying coveralls. Even as vehicles improved, though, NASA insisted that astronauts wear their suits during launch and reentry. American astronauts had never required the extra margin of safety that the suits provided, but in 1971, three Soviet cosmonauts aboard Soyuz 11 asphyxiated during reentry when an external valve failed to seal.[70] The men had not been wearing pressure suits at the time, which Soviet space authorities had considered an unnecessary redundancy (and which could not have fit easily inside the cramped reentry module).[71]

While NASA was developing a sophisticated shuttle suit for EVA, it did not intend its crews to need one for routine flights; the shuttle was the first American space vehicle to provide an oxygen-nitrogen environment at a pressure comparable to that found on Earth. Even routine shuttle operations demanded some emergency protection for the crew, though, and for the first shuttle flights, NASA found itself, oddly, without a spacesuit designed for the task, adopting one developed for pilots of Lockheed's high-altitude reconnaissance aircraft, the SR-71A. Intended as an emergency system for flying at altitudes of 85,000 feet and lacking the thermal and micrometeoroid resistance required for space environments, the intravehicular suits would protect crews only inside the shuttle's cabin. Satisfied with the reliability of the shuttle during its first four flights, though, NASA, in late 1982, abandoned the routine use of even these pressure suits by flight crews, sending

Figure 4.5
Challenger astronauts Commander Francis "Dick" Scobee (front right), Pilot Michael Smith (front left), and Mission Specialists Judith Resnik (rear right) and Ellison Onizuka (rear left) train for the STS-51L in a mission simulator. The photograph, taken in front of the cockpit, shows the crew position at launch and landing; the remaining crew members sat in chairs one deck below. All seven crew members wore the crash helmets and jumpsuits shown here; none wore a pressure suit or pressurized helmet. NASA photo by Bill Bowers.

shuttle astronauts aloft for the next three years in jumpsuits and crash helmets supplemented by portable oxygen systems inadequate for survival except at relatively low altitudes (figure 4.5). NASA's decision, though foolhardy in retrospect, was ultimately of little consequence, given the vulnerabilities in the shuttle's design.

While spacesuits provided at least marginal protection to flight crews, the real danger to the shuttle lay in its design, from its hazardous launch to its unforgiving aerodynamics upon landing. Ironically, the shuttle's dual flight modes—as rocket and as airplane—worried the pilots who flew it instead of encouraging them. The shuttle, ultimately, performed neither function well; maintaining proficiency in its peculiar gliding mode was particularly challenging. Even seasoned pilots found the shuttle a hazard to land and worried constantly about crashing it. Ironically the craft built as a "pilot's spacecraft" proved unpleasant to fly—an accommodation to the divergent constituencies whose needs turned Max Faget's original DC-3 double-flyback orbiter into a somewhat ungainly mess.

Unlike conventional aircraft, the shuttle was more glider than airplane, and a heavy, fast one at that: two qualities detrimental to handling. Only during landing did the orbiters require conventional aircraft-style piloting by the shuttle's commander and pilot, but this period was among the most perilous for the craft, and control of the vehicle during this window was left to the more experienced commander.[72] Landing required an unpowered, gliding approach (referred to as a "dead-stick landing"), offering just a single chance at safe touchdown, eased only by the size of the runway for safety. During the early 1980s, NASA lobbied for an automatic landing system that required manual piloting only during the last few seconds of touchdown; tested on STS-3, the system proved even more hazardous, giving Commander Jack Lousma (a Skylab veteran who had never flown the shuttle) only seconds to familiarize himself with the shuttle's handling characteristics before hitting the ground.[73] (A fully automatic landing system later installed on the shuttles was never used in flight.) As shuttle pilots were typically promoted to commander after only a single flight, the pilot's lack of responsibility during landing made the shuttle a poor training tool; on Columbia's maiden flight, Crippen recounted that his primary responsibility was to ensure that the landing gear deployed prior to touchdown.[74] Eventually, NASA instituted a brief period during landing when the commander turned control of the shuttle over to the pilot, so that the pilot would have a least some experience flying the shuttle before assigned to land the vehicle as commander of a future flight.[75]

The landing strip at KSC, a cost-saving measure that reduced turnaround time and transportation costs that stemmed from flying newly returned orbiters from the landing strip at Edwards Air Force Base, also proved unpopular with astronauts and fell into disuse. Edwards was built on a dry lakebed that could accommodate emergency landings; though of ample length to support shuttle landings, the Kennedy strip was surrounded by swamps that would doom an already stricken shuttle if it missed the runway or rolled off it. "I often joke," Crippen noted, "that they've got a fifteen-thousand-foot runway, but they built this moat around it and filled it full of alligators to give you an incentive to stay on the runway."[76] The shuttle's proposed second launch site at Vandenberg Air Force Base in California, intended for polar launches of reconnaissance satellites, revealed another technical hiccup that threatened the rationale for the entire program. Problems with using the shuttle for polar launches emerged almost immediately when the shuttle's gross weight began to climb during production. Too heavy to reach polar orbit, the shuttle required further modifications to the SRBs to lighten them.

Other astronauts stewed over the shuttle's payload manifest; on many missions, the craft's cargo bay was literally filled with explosives, including solid propellant motors and worse. In early 1986, the by-then veteran shuttle commander Crippen found himself and his crew preparing to fly an orbiter that would carry, in its cargo bay, a liquid-fueled rocket stage. In the past, these upper stages sat atop expendable rockets to blast scientific probes into deep space. Their cryogenic propellants would not only vent combustible fumes, but the tanks themselves were certain to explode violently in the event of an accident. Shuttle program managers and astronauts had shared the same fears of such a mission for years, but flight preparations proceeded despite the obvious risks. Astronauts even took to calling the shuttle the "Death Star" and the mission "Death Star One" (after the ill-fated space station from the 1977 film *Star Wars*), wondering if the crew would ever reach Earth orbit alive.[77] Those men survived, though, in part because the mission never flew.

NASA had expected that the first years of the shuttle fleet's operation would carry below-capacity crews and follow a launch schedule of only a few flights per year. Once it had proved itself, though, though, routine operations would pack the shuttle's flight decks to capacity with crews that might launch every two weeks. Engines, thermal tiles, and computer systems, though, all proved more problematic than expected. Orbiters sometimes underwent physical modification between flights, and NASA declared the STS ready for operational missions (including commercial satellite delivery), even though each new orbiter was unique and subject to constant repairs and upgrades. The shuttle's design problems were too numerous to correct without a massive project funded by Congress: when confronted with an unsustainable program, NASA slowed the launch rate and concentrated on maintenance and refurbishment. Doing so, though, undermined the shuttle's launch frequency, economics, and political support.

To pick up the slack in the tempo of operations, NASA crammed each flight with as many on-orbit activities as possible.[78] Military satellite deployment and on-orbit reconnaissance research, for example, soon occurred alongside scientific and commercial activities. Beginning with STS-4, three flights prior to 1986 carried military or classified intelligence payloads; NASA restricted the crews for these missions to astronauts with military commissions, and new security procedures during training and flight greatly complicated normally routine operations.[79] An Air Force program to recruit and train its military Payload Specialists—dubbed Manned Spaceflight Engineers (MSEs)—was also well underway, with the program having selected the first thirteen MSEs in 1979, another fourteen in 1982, and five in 1985.[80]

On STS-5, the first operational shuttle, Columbia, deployed two commercial communication satellites and flew the first NASA scientist-astronauts since 1974, Allen and Lenoir. NASA, though, seemed little prepared for the scientific aspects of the flight, providing the men with crew seats but so little in the way of actual duties that Allen joked to the press that NASA had placed him "'in charge of religious activities'" for the flight.[81] The mission's success, though, paved the way for additional ones, including STS-6's deployment of the first satellite of the Tracking and Data Relay Satellite System (TDRSS), a constellation of repeaters that enabled continuous contact between shuttle crews and ground controllers,[82] and the first launch of Spacelab on STS-9. Until the launch of TDRSS, NASA could maintain radio contact only when the shuttle moved in the range of ground-based dish antennas, making routine communications and scientific work difficult.

This period in the shuttle's operational history, stretching from 1982 to 1986, constituted a critical era in the history of human spaceflight despite its ultimate brevity. Attempting to demonstrate the shuttle's value as both satellite delivery vehicle and a scientific tool, NASA flew it as frequently as it could, amid popular expectations that routine, inexpensive access to space—and even a kind of democratization of the NASA astronaut corps—was just over the horizon. In offering this vision of a forthcoming Golden Age of space exploration, NASA employed the same mix of upbeat copywriting and high-quality artistic promotion that had fueled public fascination with (and congressional approval of) Apollo. The NASA Art Program, the creation of the legendary NASA administrator James Webb, had brought some of the nation's leading artists to NASA facilities in the 1960s to sketch and paint space vehicles, ground crews, and astronauts. Reassembled by NASA's Robert Schulman in 1977 to depict the space shuttle, the Art Program brought a mixture of famous and less well known artists to view the astronaut training and Columbia's launch and landing in 1981.[83] (Art reflecting NASA's secret coordination with the Air Force and National Reconnaissance Office was absent, despite persistent public speculation about the program's military role.)

NASA unveiled the dozens of works prepared for public view in late 1982.[84] While most artists approached their subjects with a romantic, vaguely impressionistic style, Robert Rauschenberg's lithograph *Hot Shot* blended photographs and schematics of a shuttle orbiter, the Personal Rescue Enclosure (PRE), and various Florida billboards announcing "Shuttlemania" in a signature-style collage that simultaneously celebrated the technology and expressed a certain measure of sarcasm about it, much as Rauschenberg had done with his "Stoned Moon" series of prints from 1969–1970 (figure 4.6).[85] Accompanying notes that he drafted to explain the work struck an upbeat tone declaring the artist's "BELIEF IN THE SPIRITUAL AND PHYSICAL IMPROVEMENT OF THE LIFE AND MIND THROUGH SPACE CURIOSITY." Rauschenberg, though, had edited the original draft statement by hand, crossing off a qualifier to his statement "~~SUCCESSFULLY OR NOT~~,"[86] a curious addition not referenced by NASA later in its promotion of the work (figure 4.7).[87]

A 1983 NASA film documentary titled *We Deliver* chronicled the achievements of this fleeting period, describing successful satellite deployments, pharmaceutical research, materials processing experiments, and the orbiting of America's first female and African American astronauts. Among the most promoted features of the shuttle (referenced in Rauschenberg's painting) was the ease with which spacesuited crew members could leave the orbiters to conduct spacewalks (EVA). This feature, too, was greatly exaggerated and premised on a belief that future work in space required equipping individual workers to do it. That future automation might make EVA unnecessary was never part of the shuttle concept; instead, the shuttle perpetuated a model of space labor that computers and robotics had already rendered obsolete.

Early American spacecraft had, like those of Project Mercury, either lacked the capacity to support EVA or, like Projects Gemini and Apollo, required that crews depressurize the entire space vehicle in order for a crew member to exit. Once outside, astronauts had found maneuvering in the weightlessness of orbit to be extremely taxing with the spacesuit technologies of the day, and even in the Moon's reduced gravity field, the notion of working while spacesuited required significant adjustments in movement and expectations. An inflated balloon kept from expanding by netting, cables, and synthetic fabrics, the spacesuit exerted pressure against its wearer's movements and offered a limited range of motion. With limbs and digits constrained

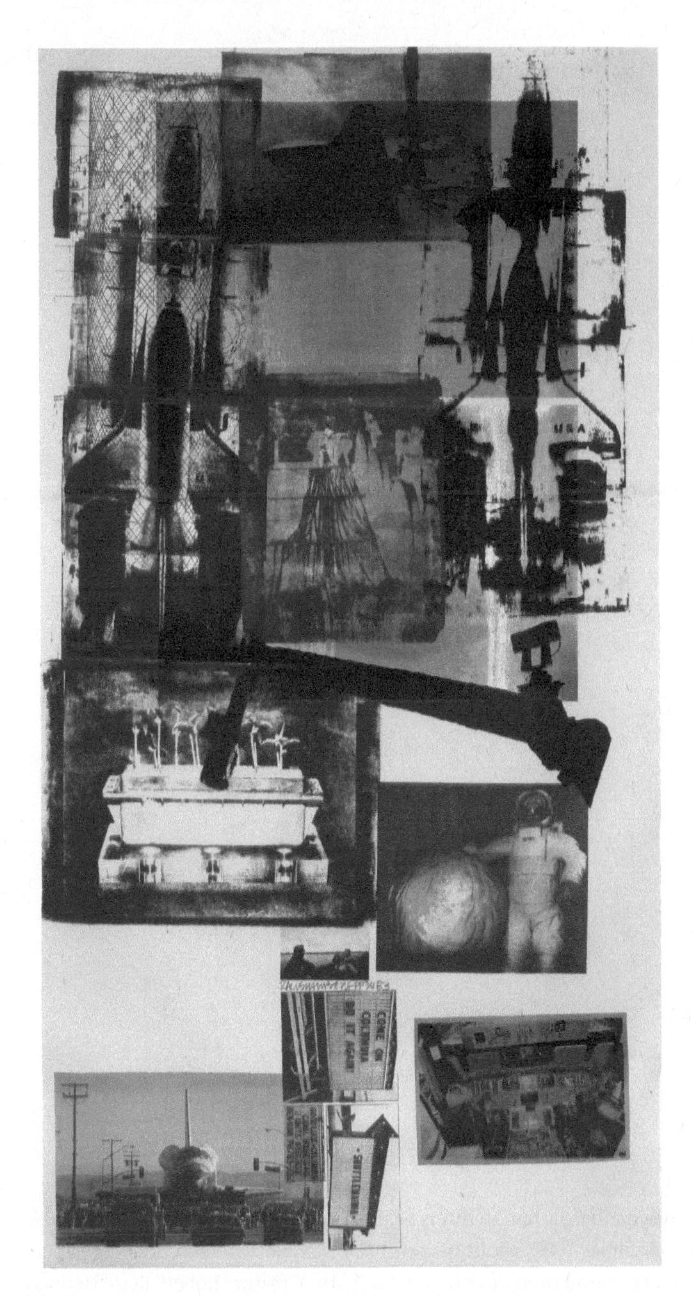

Figure 4.6
Robert Rauschenberg (American; 1925–2008). *Hot Shot*, 1983. Lithograph in nineteen colors on Arches Cover buff paper 81×42 in. (205.7×106.7 cm). Williams College Museum of Art, Williamstown, MA: Gift of Hiram Butler, MA '79 in honor of Earl A. Powell III, Class of '66 (M.2016.9). (From an edition of 29, published by Universal Limited Art Editions, West Islip, New York. Copyright is held by the Robert Rauschenberg Foundation.)

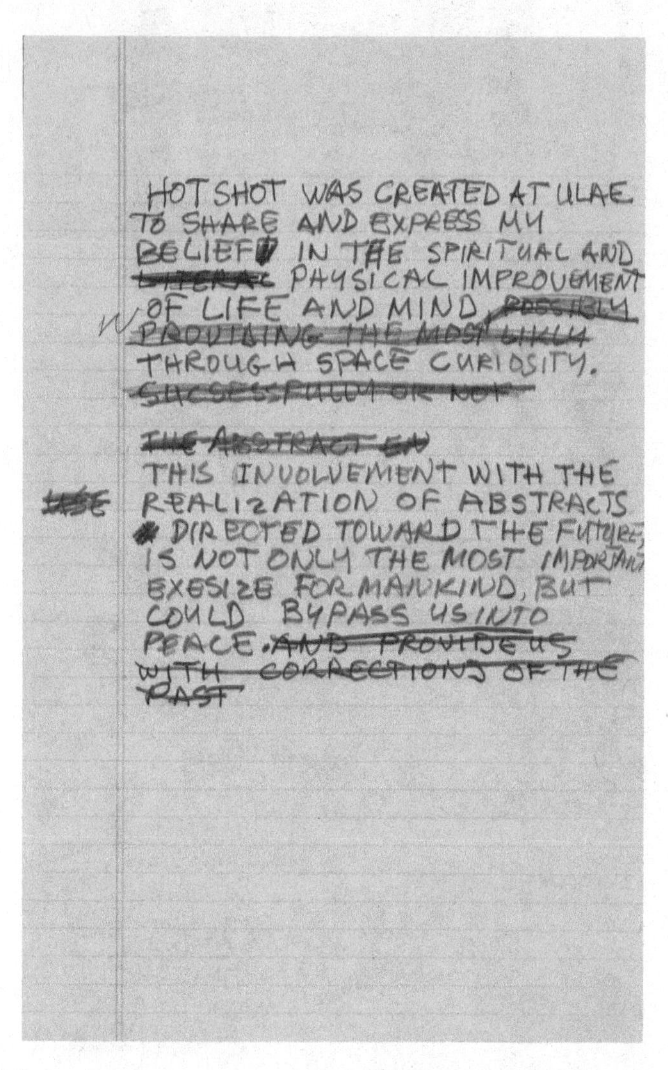

HOT SHOT WAS CREATED AT ULAE
TO SHARE AND EXPRESS MY
BELIEF IN THE SPIRITUAL AND
~~LITERAL~~ PHYSICAL IMPROVEMENT
OF LIFE AND MIND, ~~POSSIBLY~~
~~PROVIDING THE MOST LIKLY~~
THROUGH SPACE CURIOSITY.
~~SUCSESSFULLY OR NOT~~

~~THE ABSTRACT EN~~
THIS INVOLVEMENT WITH THE
REALIZATION OF ABSTRACTS
DIRECTED TOWARD THE FUTURE,
IS NOT ONLY THE MOST IMPORTANT
EXESIZE FOR MANKIND, BUT
COULD BYPASS US INTO
PEACE. ~~AND PROVIDE US~~
~~WITH CORRECTIONS OF THE~~
~~PAST~~

Figure 4.7

Robert Rauschenberg's handwritten draft of a statement for *Discover* magazine, published in December 1982, about his print *Hot Shot*. Robert Rauschenberg Papers. Robert Rauschenberg Foundation Archives, New York. Creator: Robert Rauschenberg, 1982. (Copyright is held by the Robert Rauschenberg Foundation.)

by layers of rubber, nylon, and fiberglass, few points outside a spacecraft to anchor the body in weightless conditions, and the constant drift of tools and equipment, astronauts attempting to spacewalk experienced dehydration, overheating, and elevated heart rates so severe that NASA flight surgeons cut their sojourns short. These difficulties were cause for alarm during the 1960s, as most plans for future space activities revolved around individual human access to space, either on the lunar surface or assembling space stations in Earth orbit, a goal of planners like Wernher von Braun since the 1950s. Indeed, early plans for the large, toroidal space stations that he advocated envisioned astronauts in spacesuits or miniaturized spacecraft assembling stations by hand, like cosmic construction crews.[88]

Improved training methods, new suit technologies, and a better understanding of the limitations of the space environment during Project Gemini solved many of these problems before Apollo's Moon landings, but a number of problems remained. Experiments aboard Skylab in the mid-1970s addressed some of the remaining concerns, including safe egress and ingress from a spacecraft and maneuvering capability, but only partially. Large enough to accommodate multiple pressurized chambers, Skylab featured an airlock and a large, openable hatch derived from Gemini.[89] The station's large interior volume also permitted the testing of an experimental Air Force jet pack powered by pressurized nitrogen gas—the Astronaut Maneuvering Unit (AMU)—that would enable astronauts to move more effectively; while never tested outside the Skylab Orbital Workshop, its use within it verified the principle of its operations. (An earlier test of this pack in the 1966 Gemini IX-A mission ended in failure when the spacewalking astronaut attempting to retrieve the AMU from the rear of the spacecraft overheated while making his way aft.)[90] While these exercises suggested the practicality of future orbital operations, spacewalks remained relatively rare and risky.

The space shuttle, designed with its own EVA airlock nestled in the crew compartment, would, its designers hoped, demonstrate the maturity of EVA technologies by enabling crew members to retrieve and repair satellites and construct space stations in orbit as von Braun had predicted (figure 4.8). The shuttle's Extravehicular Mobility Unit (EMU), an improvement on the Apollo EVA spacesuit, offered long-duration comfort and faster donning and removal; as early as possible, NASA would test both the EMU and its new jet pack, the Manned Maneuvering Unit (MMU) in space, demonstrating their utility for on-orbit repair and construction work. In February 1984,

Figure 4.8
STS-51A Mission Specialist Dale Gardner appears in a photograph taken by fellow Discovery mission specialist Joe Allen after recovering two stranded communications satellites for recovery and refurbishment in November 1984. The photograph attracted controversy by revealing that the role of NASA's space shuttle was principally commercial. NASA photo.

Mission Specialist Bruce McCandless achieved the first untethered spacewalk in Earth orbit using the EMU and MMU. That mission, STS-41B (named using NASA's complex new mission numbering scheme), also saw the deployment of two communications satellites whose upper stages failed, leaving them intact but stranded in orbits too low to be useful. Later that year, astronauts James van Hoften and George Nelson attempted to use the MMU to recover and repair a malfunctioning satellite designed to study solar flares but failed in the effort. (The Canadarm Remote Manipulator System recovered the satellite instead.) And in November 1984, the STS-51A astronauts Allen and Dale Gardner recovered the two stranded communications satellites launched earlier that year and returned them to Earth in the shuttle's cargo bay for refurbishment and later relaunch.

Actual efforts to use the MMU offered a mix of promise and concern. The apparatus worked as designed, but the servicing of satellites in space proved extremely time-consuming, with procedures that might take minutes on Earth taking hours in space. NASA soon retired the bulky unit in favor of smaller, less capable maneuvering devices, and as schedule slippage mounted, the on-orbit repair missions requiring them became less frequent. The astronauts' continued efforts to utilize jet-pack technology in spite of its dangers serve as an example of both their dedication to mastering a technology and the anachronism of the technology that they mastered. Although work processes on Earth had automated substantially between the 1950s and the 1980s, NASA's vision of orbital activity had remained virtually static in this time period, with promotional materials still imagining hand assembly of large, sophisticated structures by an army of highly trained astronauts.[91] Most EVA technology of the 1960s, though, was brief and experimental, and operational activities in the 1970s were limited to the installation of instrumentation and the retrieval of lunar samples and film cartridges: important jobs, but ones easily automated in future craft (which would dispense with film cameras entirely in favor of digital devices). Computerized autopilots employing radar could dock space vehicles more safely than could humans attempting to link them by hand.

While NASA persisted in its belief that EVA was the key to future space operations, history seemed to offer striking counterevidence: every space station orbited in the 1970s (the Soviet Salyut and Almaz and the American Skylab) was launched intact on a large rocket rather than assembled from parts. Even when orbital assembly techniques were later employed on the Russian Mir and the International Space Station, they were limited to automated rendezvous and docking of large craft, not hand assembly. While certain missions included external repairs to the stations (most notably by the 1973 Skylab II crew), EVA suits remained dangerous instrumentalities prone to breakage, and human labor in space remained extremely time-consuming. Astronauts working outside the craft for a full business day or longer had to eat, drink, and excrete in their suits. Although astronauts were weightless, motions that were effortless on Earth often took considerable strength, and astronauts often lost fingernails due to repetitive stress injuries of their fingertips. Not only were humans physically vulnerable in space, but as it as it turned out, they were also an overpriced and contentious labor force (as labor unrest aboard Skylab demonstrated in the Skylab IV crew's "strike in space").[92]

With automation technologies advancing rapidly, the shuttle's continued reliance on human activity and human control became a point of pride, a justification for continued human spaceflight initiatives, and a source of inefficiency: most missions intended for the shuttle (like satellite and space station delivery to orbit) could be conducted remotely by automated craft, and the few activities (like satellite repair) that required active human intervention were potentially so dangerous and difficult that their success might not justify their expense. (Simply launching replacement satellites cost barely more than retrieving and repairing malfunctioning ones, which also risked human lives and a space shuttle in the process.) Proving that orbital repair *could* be done was not evidence that it *should* be, but NASA expected that, in time, better technologies and methods would make the space environment amenable to human labor rather than make such activities obsolete.

Central to this hope was the belief that the imminent routinization of spaceflight created by the space shuttle would make space repair more frequent and construction less expensive. NASA's experience of the shuttle through 1985 offered neither, but the pace of flights was increasing, and emboldened by these successes, the agency hoped that forthcoming improvements would resolve the shuttle's issues. With new orbiters and enough launch and landing facilities, the shuttle program might even maintain a schedule that afforded every member of the astronaut corps of between 130 and 190 men and women the opportunity to fly at least once a year, while creating opportunities for noncareer passengers to fly as well.[93] This latter expectation was an early goal of the STS and a key element of NASA's sales pitch for it: the willingness to fly nonprofessional tourists would signal the shuttle's safety and create more paying business of the kind that NASA actually wanted.

In the years that followed the shuttle's authorization in 1972, NASA appeared to make little headway in providing opportunities for nonprofessionals to fly on the shuttle, and institutional resistance to the idea at NASA field centers was occasionally bitter.[94] By 1983, with a small number of operational shuttle flights completed, NASA announced a formal program to recruit private citizens as crew members. From the outset, it was clear that the individuals chosen for what NASA called the Space Flight Participant Program would not serve as active members of the flight crew (in any capacity other than to, as NASA administrator James Beggs described, "tend the galley"). Rather, the agency would hope to fly "artists" and other "professional communicators" aboard the space shuttle, who would, at the conclusion of their missions, help the agency connect to an American public still indifferent to the space shuttle program.[95]

The lack of actual planning for such passengers at the time reveals the extent to which NASA's promises were puffery: the shuttle had no proper galley or passenger deck offering easy ingress: access even to the middeck seats used for scientist crew members required climbing sideways through ladders onto reclined seats. All shuttle astronauts, furthermore, required extensive training in the use of rescue and recovery gear, helmets, and escape systems. The presence even of partially trained Payload Specialists among the crew was galling to career astronauts, and the passengers themselves were often uncomfortable: despite its large volume, very little of the shuttle orbiter was habitable, amounting to little more than a 300-square-foot studio apartment for a typical crew of seven people, with workspaces doubling as communal sleeping, eating, and hygiene facilities. While it was true that the shuttle's comparatively gentle launch and reentry profiles would expose its passengers to forces that any reasonably healthy person could endure, most would probably vomit uncontrollably for at least the first day or so in orbit due to space adaptation syndrome.[96] Like promises that shuttles would visit an as-yet unbuilt space station or retrieve and repair future satellites, the claim that it would carry large numbers of everyday people into space was fantasy. And the odds than any one citizen would fly the shuttle were astronomical, as NASA could offer even rudimentary training to only a handful of people at a time.

Previous discussions during the 1970s had raised the possibility of orbiting of journalists and artists aboard NASA vehicles;[97] such flights had precedents in wartime, as US military forces had frequently embedded writers and artists within combat units. These individuals who were chosen performed a communication role perceived as essential to national morale and possessed skills that average members of the public lacked. In considering candidates for the Spaceflight Participant program, though, President Reagan hoped to select crew members who would inspire children, and he also wanted to use the selection to, during a vigorous reelection campaign, salvage the votes of educators, who had condemned his administration for cuts in federal educational funding.[98] NASA's invitation in August 1984 to fly the first private citizen into space, thus, extended the opportunity, not to a journalist, but to an elementary or secondary school teacher.[99] Reagan's choice of a teacher reflected a somewhat old-fashioned view of youth culture, and NASA's public relations campaign for the selection stripped the project of any semblance of adventure. At the news conference announcing the "Teacher-in Space" program, Beggs described the shuttle a "'benign, shirt-sleeved environment' that 'allows a reasonably healthy person to fly there with nothing more than relatively rudimentary training.'"[100]

In November 1984, while selection for the Teacher-in Space program was underway, NASA announced that it had accepted a solicitation from the Utah senator Jake Garn for a seat on an impending shuttle flight. A former naval aviator, he headed the committee overseeing NASA appropriations, and he flew in April 1985, months before NASA had even announced its Teacher-in-Space selections. Again, shuttle astronauts had little input on the decision. Instead, as Commander Karol Bobko recounted, NASA flight crew operations director George Abbey called and asked him, while preparing for his upcoming mission, "'What sort of training program would you have if you had a new passenger that was only going to have eight or twelve weeks?'" Bobko, incredulous, asked why Abbey wanted to know; Abbey then broke the news that Bobko, set to command Discovery's STS-51D flight in 1985, had "'a new passenger'": Garn.[101] A Saudi prince and air force pilot, Sultan bin Salman bin Abdulaziz Al Saud, also flew the shuttle as a Payload Specialist in June 1985; the Florida Democratic congressman Bill Nelson followed Garn into space in January 1986.[102]

Meanwhile, shuttle missions during this period became overloaded with both commercial payloads and scientific objectives. Upon hearing that a NASA spokesperson had quoted a member of the 1984 STS-51A crew saying that they would recover two satellites on their mission, the mission's commander, Rick Hauck, was incensed, describing, according to crewmate Allen, the optimistic prediction as a "'fucking miracle.'"[103] The successful recovery of a single satellite was more likely.[104] Although under pressure to launch quickly, NASA, by the end of 1985, was still struggling to make twelve flights a year, with no indication that the tempo of operations would increase any time soon.[105] "The whole system was starting to crater," Hank Hartsfield later recounted, with scheduled flights hurriedly prepared so that NASA could meet its self-imposed guidelines.[106] Unable to achieve the economies of scale sought, the shuttle, by the end of 1985, looked less and less like a commercial satellite launcher than another in a long line of experimental space vehicles.[107]

Despite the promise to broaden participation in the space program, NASA's selection for the Teacher-In-Space program in 1985 proved its most complex and discriminating. Flooded with 40,000 elementary, middle, and high school teachers expressing interest (postsecondary instructors were not eligible to apply),[108] NASA dispatched lengthy applications and requests for supporting materials, receiving 11,000 completed applications. NASA forwarded these to the education departments of the candidates' respective states or territories, which chose two from each and forwarded the list of 114 semifinalists to NASA. Interviews with an eclectic panel of experts (including the professional basketball player Wes Unseld and the actor Pam Dawber of the extraterrestrial ABC television comedy *Mork & Mindy*) narrowed this group to 10 finalists. After a photograph session with Reagan in Washington, D.C., the finalists endured a battery of medical and psychological tests at JSC and a sightseeing excursion at a theme park operated by the State of Alabama, the US Space & Rocket Center, in Huntsville. During the visit, a park employee suffered a fatal accident while attempting to impress the audience with a stunt, ending the festivities. In July 1985, NASA selected the New Hampshire social studies teacher Christa McAuliffe as a prime crew member for a future shuttle flight, with the Idaho second-grade teacher Barbara Morgan serving as her backup.

Unlike with previous astronaut assignments, NASA assigned McAuliffe to a single mission for which the flight crew had already been selected; the crew, led by Commander Dick Scobee (a former lifting body test pilot who joined NASA in 1978),[109] greeted her upon her return to JSC. Without piloting or scientific duties on the forthcoming flight, McAuliffe would deliver two televised lessons for children from space.[110] According to news accounts, Scobee welcomed McAuliffe with the comment that "no matter what happens on the mission, it's going to be known as the teacher mission," a statement intended to flatter her, but which instead reflected Scobee's recognition that the public cared little for the difficult work of the flight, which would include a satellite deployment and astronomical experiments.

McAuliffe, alone among the finalists, had vomited on the rides at Huntsville; her seeming frailty endeared her to some NASA personnel and worried others.[111] Career astronauts, McAuliffe believed, resented both her presence and NASA's suggestion that she would "'humanize'" spaceflight, a remark that (perhaps inadvertently) branded her fellow crew members as inhuman. NASA's in-house educational personnel, meanwhile, were insulted that McAuliffe was chosen to fly on an educational mission despite having no NASA background, no scientific education, and no experience teaching technical subjects.[112] Indeed, she struggled to master the material for her scheduled fifteen-minute lessons to students; chosen for her personality rather than her expertise, McAuliffe filled her days with media appearances and concealed her confusion behind a cheery disposition.[113]

Despite claiming publicly to have been accepted by her fellow astronauts, McAuliffe was often treated as a dangerous amateur let loose on the shuttle; during training for her flight, Scobee instructed McAuliffe not to touch any of the switches, even in the shuttle simulator at JSC.[114] The space shuttle, despite statements otherwise, had not been designed for passengers and could be destroyed relatively easily through the careless operation of its many controls. Astronauts were aware of these vulnerabilities, but "joyriding" "part-timers" (as NASA career astronauts called them)[115] were not. "Individuals who were clueless about the risks of spaceflight were being exploited for public relations purposes," later wrote Mullane. "The entire part-timer program was built on the lie that the shuttle was nothing more than an airliner."[116]

McAuliffe's flight aboard Challenger, along with the six astronauts who would shepherd her into space, would occur in late January 1986.

Conclusion

In November 1985, an article in *Popular Mechanics* by Dennis Eskow entitled "Space Vacation 1995" predicted that within a decade, America's newest spacecraft—the space shuttle—would accommodate twenty-four paying passengers in a commercially designed compartment containing seats, a bar, an observation deck, and an exercise area. The space shuttle, at the time, was only a little more than four years into its flying history and NASA had only recently declared it "operational." Its earliest flights had carried only skeleton crews, and yet observers still could imagine that in ten years, the most ambitious dreams of the shuttle's creators would come true. Paying $1 million each for the privilege, the article continued, the first space tourists would be medically vetted, rudimentarily trained, and offered a three-day orbital voyage. Simultaneously grandiose and mundane, the image of this space hotel still fascinates nearly thirty years after its imagined invention. Built on the most optimistic assumptions about the capabilities of the STS and the frequency with which NASA could launch its space shuttle fleet (an impossible twenty-four flights per year), these predictions represented one of the last examples of a decade-long wave of enthusiasm about the possibilities presented by a reusable orbital spaceplane: that it would restructure the

Table 4.1
STS astronaut assignments (1981–1986)

Year	Mission Number			
	Prime Crew (NASA Selection Year) (Role)			
1981	STS-1		STS-2	
	John Young ('62) (CDR)		Joe Engle ('66) (CDR)	
	Bob Crippen ('69) (PLT)		Dick Truly ('69) (PLT)	
1982	STS-3	STS-4 (Military Flight)		STS-5
	Jack Lousma ('66) (CDR)	Ken Mattingly ('66) (CDR)		Vance Brand ('66) (CDR)
	Gordon Fullerton ('69) (PLT)	Hank Hartsfield ('69) (PLT)		Bob Overmyer ('69) (PLT)
				Joe Allen ('67) (MS1)
				Bill Lenoir ('67) (MS2)
1983	STS-6		STS-7	
	Paul Weitz ('66) (CDR)		Bob Crippen ('69) (CDR)	
	Karol Bobko ('69) (PLT)		Rick Hauck ('78) (PLT)	
	Story Musgrave ('67) (MS1)		John Fabian ('78) (MS1)	
	Don Peterson ('69) (MS2)		Sally Ride ('78) (MS2)	
			Norm Thagard ('78) (MS3)	
	STS-8		STS-9	
	Dick Truly ('69) (CDR)		John Young ('62) (CDR)	
	Dan Brandenstein ('78) (PLT)		Brewster Shaw ('78) (PLT)	
	Guy Bluford ('78) (MS1)		Own Garriott ('65) (MS1)	

Table 4.1 (*continued*)

STS astronaut assignments (1981–1986)

Year	Mission Number			
	Prime Crew (NASA Selection Year) (Role)			
	Dale Gardner ('78) (MS2)		Robert Parker ('67) (MS2)	
	Bill Thorton ('67) (MS3)		Ulf Merbold (PS1) (ESA)	
			Byron Lichtenberg (PS2)	
1984	STS-41B	STS-41C		STS-41D
	Vance Brand ('66) (CDR)	Bob Crippen ('69) (CDR)		Hank Hartsfield ('69) (CDR)
	Hoot Gibson ('78) (PLT)	Dick Scobee ('78) (PLT)		Mike Coats ('78) (PLT)
	Bruce McCandless II ('66) (MS1)	Terry Hart ('78) (MS1)		Mike Mullane ('78) (MS1)
	Ronald McNair ('78) (MS2)	James van Hoften ('78) (MS2)		Steven Hawley ('78) (MS2)
	Robert Stewart ('78) (MS3)	Pinky Nelson ('78) (MS3)		Judith Resnik ('78) (MS3)
				Charlie Walker (PS1)
	STS-41G		STS-51A	
	Bob Crippen ('69) (CDR)		Rick Hauck ('78) (CDR)	
	Jon McBride ('78) (PLT)		David Walker ('78) (PLT)	
	Kathryn Sullivan ('78) (MS1)		Joe Allen IV ('67) (MS1)	
	Sally Ride ('78) (MS2)		Anna Fisher ('78) (MS2)	
	David Leestma ('80) (MS3)		Dale Gardner ('78) (MS3)	
	Paul Scully-Power (PS1)			
	Marc Garneau (CSA) (PS2)			

STS-51C (Military Flight)
Ken Mattingly II ('66) (CDR)
Loren Shriver ('78) (PLT)
Ellison Onizuka ('78) (MS1)
James Buchli ('78) (MS1)
Gary Payton (MSE) (PS1)

STS-51G
Dan Brandenstein ('78) (CDR)
John Creighton ('78) (PLT)
John Fabian ('78) (MS1)
Steve Nagel ('78) (MS2)
Shannon Lucid ('78) (MS3)
Patrick Baudry (CNES) (PS1)
Salman Al-Saud (RSAF) (PS2)

STS-51J (Military Flight)
Karol Bobko ('69) (CDR)
Ronald Grabe ('80) (PLT)
David Hilmers ('80) (MS1)
Robert Stewart ('78) (MS2)
William Pailes (MSE) (PS1)

STS-51D
Karol Bobko ('69) (CDR)
Donald Williams ('78) (PLT)
Rhea Seddon ('78) (MS1)
David Griggs ('78) (MS2)
Jeffrey Hoffman ('78) (MS3)
Charlie Walker ('78) (PS1)
Jake Garn (PS2)

STS-51F
Gordon Fullerton ('69) (CDR)
Roy Bridges, Jr. ('80) (PLT)
Karl Henize ('67) (MS1)
Story Musgrave ('67) (MS2)
Anthony England ('67) (MS2)
Loren Acton (PS1)
John-David Bartoe (PS2)

STS-61A
Hank Hartsfield ('69) (CDR)
Steven Nagel ('78) (PLT)
Bonnie Dunbar ('81) (MS1)
James Buchli ('78) (MS2)
Guy Bluford ('78) (MS3)
Reinhard Furrer (DLR) (PS1)

STS-51B
Bob Overmyer ('69) (CDR)
Fred Gregory ('78) (PLT)
Don Lind ('66) (MS1)
Norm Thagard ('78) (MS2)
Bill Thorton ('67) (MS3)
Lodewijk van den Berg (PS1)
Taylor Wang (PS2)

STS-51I
Joe Engle ('66) (CDR)
Richard Covey ('78) (PLT)
James van Hoften ('78) (MS1)
John Lounge ('80) (MS2)
William Fisher ('78) (MS3)

STS-61B
Brewster Shaw ('78) (CDR)
Bryan O'Connor ('80) (PLT)
Jerry Ross ('80) (MS1)
Mary Cleave ('80) (MS2)
Sherwood Spring ('80) (MS3)
Charles Walker ('78) (PS1)

Table 4.1 (*continued*)
STS astronaut assignments (1981–1986)

Year	Mission Number			
	Prime Crew (NASA Selection Year) (Role)			
		Ernst Messerschmid (DLR) (PS2)		Rodolfo Vela (PS2)
		Wuddo Ockels (ESA) (PS3)		
	STS-61C		STS-51L	
	Hoot Gibson ('78) (CDR)		Francis Scobee ('78) (CDR)	
	Charles Bolden (PLT)		Michael Smith ('80) (PLT)	
	George Nelson ('78) (MS1)		Ellison Onizuka ('78) (MS1)	
	Steven Hawley ('78) (MS2)		Judith Resnik ('78) (MS2)	
	Franklin Chang Diaz ('80) (MS3)		Ronald McNair ('78) (MS3)	
	Bill Nelson (PS1)		Gregory Jarvis (PS1)	
	Robert Cenker (PS2)		Christa McAuliffe (PS2)	

Abbreviations:
CDR: Commander
PLT: Pilot
MS: Mission Specialist
PS: Payload Specialist
CSA: Canadian Space Agency
CNES: Centre National D'études Spatiales (National Centre for Space Studies, France)
DLR: Deutsches Zentrum für Luft- und Raumfahrt e.V. (German Aerospace Center)
ESA: European Space Agency
MSE: Manned Spaceflight Engineer (US Department of Defense)
RSAF: Royal Saudi Air Force

economics of space exploration, making it more akin to airline travel than a polar expedition.

The *Popular Mechanics* article presented an intriguing but improbable vision of the future. NASA had no plans to fly tourists on the space shuttle, and no private entities had reached an agreement with the shuttle's prime contractor, Rockwell International, to manufacture additional shuttles for recreational use.[117] Early shuttle flights revealed a space infrastructure requiring extensive maintenance and much higher per-mission costs than anticipated. And the onboard experience of the shuttle offered cramped quarters and unpalatable food, with no plans to augment it with hotel-like accommodations. None of the promised amenities, strictly speaking, were ever expected for the craft.

As originally envisioned, NASA's space shuttle was a single component of a constellation of spacecraft, space stations, and interplanetary vehicles, only some of which might offer anything close to comfortable accommodations. Thin 1970s budgets could not support this infrastructure, so NASA pitched the shuttle instead as an inexpensive, reusable, stand-alone craft with enough extra seats to carry passengers, including the women and minority personnel who had been ineligible for previous astronaut selections due to their lack of military flying experience. While NASA had begun a tentative program to occasionally offer shuttle seats to well-connected politicians and even a schoolteacher, no space hotels were on the horizon. When the Eskow article appeared, in fact, NASA was slowly moving the shuttle away from a purely civilian orientation and toward military operations that would periodically swap out civilian crews for military astronauts who would use the shuttle to deploy spy satellites after launching from an alternative site in California. The use of the shuttle in this way was a significant break in precedent from NASA's earlier human spaceflight projects and a condition of its development when the Nixon administration approved the shuttle's construction in 1972.

The incompatible goals for the shuttle—commercialization and democratization of space travel, and military dominance of space—characterized not only the physical construction of the shuttle during the 1970s, but contemporaneous plans for the composition of the astronaut corps and the crewing of shuttle flights. Each promise made by the shuttle's supporters in NASA, Congress, the military, and the industry about the shuttle's capabilities pushed the vehicle further toward crisis, as its shortcomings became gradually more obvious. In ways large and small, NASA management appeared to grease the skids for the loss of Challenger by flying missions using vehicles with known anomalies, by removing rather than augmenting safety systems, and by overtaxing the shuttle's limited throughput with more and more demanding missions. Against the backdrop of these pressures, the loss of Challenger in 1986 from a predictable set of inflight problems appeared to be the inevitable result of a colossal management failure, or even a case of the inexorable danger of complex technologies biting the hands of their designers. In fact, it proved to be neither.

5 A History Rooted in Accident, 1986–2011

"The glitter has rubbed off the job of working at the Johnson Space Center," the *Washington Post* reported on February 5, 1986, but the loss of the space shuttle Challenger and its crew weeks earlier had not been the cause. Instead, the article noted that "the glamor began to fade last July when the space shuttle Challenger rode into orbit on two engines instead of three." The launch hiccup—the failure of a heat sensor that resulted in the premature automatic shutdown of one of the three Space Shuttle Main Engines (SSMEs)—was a peculiar form of emergency: serious enough to warrant special contingency measures but relatively tame in impact. In the event that one SSME failed, the shuttle would still reach Earth orbit, but lower than planned. "'Abort to orbit,' it was called at the time," the *Post* noted of the incident.[1] In a space program that had, until that point, been relatively free of well-publicized in-flight mishaps, the failure was both a troubling warning and a metaphor for the Space Transportation System (STS). Although giving the appearance of an operational infrastructure (deploying satellites and conducting scientific work), the space shuttles' flights from 1981 to 1986 were marred by a series of worrisome problems.

The problem that struck STS-51F came as little surprise to those who designed and operated the shuttle: the three main engines mounted at the rear of the orbiters operated under unprecedented pressure, using one of the most energetic propellant combinations available: liquid hydrogen and liquid oxygen, each chilled to hundreds of degrees below zero. Unlike previous engines using this combination of propellants, the SSMEs were intended to be reusable for up to 100 flights, a level of performance never expected from a propulsion system devised to work for only a few minutes. The National Aeronautics and Space Administration (NASA) had hoped that the engines would require little maintenance between flights, but as problems mounted, the engines required almost complete disassembly and recertification before each flight, an expensive process representative of the tremendous risks that NASA sought to control. On July 29, 1985, though, Challenger dodged a bullet: it had not failed, but neither had it quite succeeded. Procedures intended to compensate for problems had worked, but the abort had left the orbiter where nobody wished it to be.

Abort to Orbit (ATO) was one of several contingencies scripted by NASA in the event that the shuttle suffered a serious malfunction on launch. Astronauts, to the extent possible, trained for these aborts and carried printed procedure books in the cockpit to perform the often-complex steps required for each. Each abort mode was keyed to a failure of the shuttle's propulsion system at various points in the flight, from a relatively late, partial failure (ATO) to more crippling early failures of the main engines or solid rocket boosters (SRBs). A more profound propulsion system problem (or inflight emergency) required an Abort Once Around (AOA), in which the shuttle, having limped into too low an altitude to sustain multiple revolutions around the Earth, would reenter the atmosphere and land after a single ninety-minute circumnavigation. A more disabled shuttle unable to enter orbit could, alternatively, coast on a suborbital trajectory to Transoceanic Abort Landing (TAL) on a predesignated airstrip several thousand miles east of the launch site in Florida.

More extreme and most difficult to execute would have been a Return to Launch Site (RTLS) abort, which would have been the earliest option available to the crew. The crew would jettison the shuttle's boosters after they burned out and fire the SSMEs to try to loop the shuttle in a return trajectory that would place it on a path to land on the runway at NASA's Kennedy Space Center (KSC). Astronauts regarded this abort mode as impossible to execute: NASA had considered attempting it during an early shuttle test flight, but astronauts refused, finding the abort more dangerous to fly than an orbital mission.[2]

However unlikely to succeed, the RTLS abort was still preferable to the emergency procedures—or lack thereof—for the shuttle during the first critical moments when the SRBs were firing or following its arrival in orbit. There were no abort modes for SRB failure on launch, inflight hull breach and loss of pressurization, reentry damage to the shuttle's wings, or any problem with the shuttle's airworthiness upon landing—other than a hatch, a few parachutes, and the ejection seats provided to the two members of Columbia's flight crew. The space shuttle was, simply, only half-complete: safety systems intended for the craft were shed during its tortuous development.

By the end of January 1986, the inherent compromises in the shuttle's design and unrealistic expectations of its use had manifested themselves in a fatal accident: the loss of Challenger shortly after liftoff. Scholars, engineers, and politicians who examined the loss of Challenger (particularly the expert panel assembled by President Ronald Reagan under former secretary of state William P. Rogers) found issues in design and maintenance, but also evidence of deeper problems in NASA's management of the shuttle program, including excessive pressure to launch. In her later revisionist review of the case, the sociologist Diane Vaughan fixed upon an idea that she called the "normalization of deviance"—NASA's creeping acceptance of anomalous performance—rather than the vehicle's technical shortcomings or schedule pressures to explain why a vehicle so flawed had flown for so long.[3] A sociological investigation premised on the centrality of organizational dynamics in explaining failure, Vaughan's analysis offered a cursory description of the shuttle's development and glossed over the orbiters' known technical flaws, the unique political circumstances surrounding the launch, and questionable actions by specific individuals.

The peculiar dangers that the Challenger and Columbia astronauts experienced were not a feature of spaceflight generally. Rather, they were caused by specific flaws manufactured into a vehicle by engineers who knew about better alternatives and hoped the vehicles would never fly long enough to be undermined by the compromises they had made. The Thrust-Assisted Orbiter Shuttle (TAOS) configuration that NASA selected in 1972 lacked robust propulsion technologies, redundancies, and safety systems that would protect a crew in the event of trouble; that is what had made it cheap and why it was chosen over safer vehicles that would have cost twice as much and employed liquid-fueled boosters, escape pods, and configurations that generally offered lower risk. While NASA publicly insisted that the shuttle was safe, even the engineers and analysts who developed the TAOS configuration believed that the shuttle would be redesigned to improve its safety. Challenger's loss appeared to stem principally from the fact that these upgrades came too late; indeed, most never came at all. A second fatal accident involving the orbiter Columbia in 2003 similarly failed to derail the space shuttle program,[4] despite new evidence that the shuttle was even more fragile than the public believed. As with the Challenger disaster, NASA's management received the bulk of the blame for the accident, but as in the earlier case, the shuttle's precarious design was chiefly at fault.

The Challenger Disaster

The failure that destroyed the space shuttle orbiter Challenger on January 28, 1986, occurred almost immediately upon ignition of the shuttle's SRBs, as it had, indeed, on many previous flights of the STS (figure 5.1). The boosters, steel tubes 12 feet wide and 150 feet tall, were among the largest rockets of their kind. Each was a hollow tube packed with a rubbery propellant mixture of aluminum powder and ammonium perchlorate, with a nozzle that swiveled mechanically to direct the thrust produced by their combustion. The two motors, each weighing 1.3 million pounds and together generating half the power of the first stage of the mammoth Apollo–Saturn V Moon rocket, were too large to easily construct as unitary objects. Instead, the chemical conglomerate Morton Thiokol (the result of a merger between the rocket-maker Thiokol and the chemical firm Morton-Norwich) manufactured each

Figure 5.1
STS-51L crew members pose for their 1986 Challenger flight. Front row (left to right): Michael Smith, Dick Scobee, and Ronald E. McNair; back row (left to right): Ellison Onizuka, Christa McAuliffe, Gregory Jarvis, and Judith Resnik. NASA photo.

of the two boosters in four segments, sealed by a pair of gaskets (O-rings) held in place by tang-and-clevis joints and a series of pins. Running down the center of the propellent stack was a hollow space of varying shape into which burning propellant gases flowed before exiting through the bottom of the booster: the cross-section of an eleven-pointed star in one segment ensured a large surface area of burning material and the production of massive thrust upon ignition.

A closed tube except for the opening at its base, the booster contained the expanding gases long enough for them to exit from the rear, propelling the booster skyward. At the base of the rocket, an exhaust nozzle directed the thrust; atop the booster, a nose cone and a small compartment containing navigation and recovery equipment completed the booster. After burnout, small solid rockets ejected each booster from the spacecraft stack; each descended into the ocean via three parachutes. Once located, they were recovered by ship, disassembled, packed with propellant, and reassembled. Without complex fuel pumps or valves, the use and reuse of the SRBs were simple and inexpensive, with the boosters having as their chief flaws the fact that once ignited, they could not be throttled or shut down.

The fact that each booster was composed of a variety of joints that had to be sealed against high-pressure and high-temperature combustion gases gave rise to concerns that grew over the course of the space shuttle program.[5] Segmented SRBs were a known risk when selected for use on the space shuttle, and when used on the STS, launch stresses usually caused the booster segments to rotate at their joints, weakening them. The O-rings sealing the joints frequently failed in flight, allowing hot gases to escape from the sides of the boosters instead of their exhaust bells at the bottom.[6] While molten propellant oxides immediately sealed such breaches on most missions, colder temperatures during the launch of Challenger's STS-51L mission stiffened the O-rings, and powerful cross-winds caused more joint rotation than usual, overwhelming a makeshift plug formed by combustion products on the right booster seventy-three seconds into the flight and creating, in effect, a new rocket engine pointing sideways (figure 5.2).[7] The shuttle offered no contingency plan for an early failure of the SRBs in flight; too low for an RTLS abort, the astronauts had little hope of escaping from the supersonic craft under the circumstances.[8]

The rupture of the SRB and structural collapse of the rest of the shuttle stack occurred rapidly; once the shuttle began to break up, aerodynamic

Figure 5.2
A plume of combustion gases exits the gap in Challenger's solid rocket booster (SRB) segments moments before the destruction of the spacecraft in 1986. NASA photo.

forces tore Challenger apart, separating the armored crew compartment from the wings and tail and creating a ballistic capsule that was uncontrollable and unflyable. The astronauts likely survived the initial dismemberment of their craft, as telemetry indicates that some members of the crew (who were not wearing spacesuits, as per NASA practice at the time) activated portable oxygen systems after the event began. Without parachutes to slow its descent, the crew compartment struck the ocean at 200 miles per hour, likely killing its occupants instantly through shock rather than hypoxia or penetrating injuries (their bodies were later recovered, mostly intact).[9] Postflight analysis of film footage of the launch revealed the likely cause of the accident, and the program was placed on hold for nearly two years, pending the outcome of formal reviews.[10]

Under too much scrutiny to be handled internally by NASA, the tragedy was investigated by the Rogers Commission, loaded with celebrities (including the first man on the Moon, Neil Armstrong), but the hearings were a farce from the very beginning. Just as Nixon had chosen Rogers to serve as secretary of state due to his lack of foreign policy experience,[11] President Reagan had ordered Rogers—a political functionary with no scientific training—to spare NASA's reputation, and he dutifully ordered the commission to refrain from criticizing the agency in their deliberations.[12] Nevertheless, the investigations revealed both long-standing technical problems with the SRB joints and the extent to which managers and engineers at NASA and Morton Thiokol had debated and even stewed over the safety of the launch beforehand: Thiokol engineers raised concerns well before launch that the resiliency of the O-rings would degrade in the cold, making them more likely to fail.

NASA managers were unpersuaded at the time—even angered by the last-minute bit of bad news—and pressured Thiokol personnel to green-light the launch. Reluctantly, the contractor complied.[13] In one of the most damaging (but still somewhat ambiguous) bits of evidence, Thiokol senior vice president Jerry Mason pressured the lone holdout, an engineer named Robert Lund, to temper his criticisms, urging him to take off his "engineering hat" and put on his "management hat."[14] The Rogers Commission also encountered ample evidence that NASA management had been consumed by a desire to maintain the shuttle's breakneck launch schedule, silencing voices that urged concern.

The Rogers Commission's discoveries were particularly galling to chief astronaut John Young, commander of the first shuttle flight in 1981 and a constant behind-the-scenes critic of the shuttle's design. Managing flight crew issues related to Columbia in the late 1970s, Young frequently sparred with the shuttle program manager, Deke Slayton. Slayton, who had led NASA's Astronaut Office in the 1960s, quickly tired of Young's frequent missives—"Young grams," he called them—that called attention to various technical flaws in the vehicle, and frequently asserted that such safety consciousness would have prevented Apollo from reaching the Moon.[15]

Watching the Rogers Commission hearings on television, Young privately challenged NASA's senior management again, asserting that the flaw in the SRBs was a well-known consequence of a faulty design. Indeed, he noted, the

O-rings that sealed the segments appeared to have been designed *not* to work. Instead, they relied upon a partial failure to seal the segments properly and offered no redundancy if they did not. As he put it:

> From watching the Presidential Commission open session interviews on television [sic], it is clear that none of the direct participants have the faintest doubt that they did anything but absolutely the correct thing in launching [STS] 51-L at every step of the way. While it is difficult to believe that any humans can have such complete and total confidence, it is even more difficult to understand a management system that allows us to fly a solid rocket booster single-seal design that explosively, [and] dynamically verifies its criticality 1 performance in its application. This is because the prelaunch leak check pressurized that criticality 1 primary seal away from its proper sealing position. Sealing then relied on the single dynamic action of solid rocket motor ignition to properly seal the primary seal. The proper sealing has to be accomplished within milliseconds. If proper sealing did not occur, in Morton Thiokol Inc.'s own words, "subscale testing verified seal resiliency unable to follow gap opening in metal parts—no secondary seal activation if primary seal fails."[16]

"There is only one driving reason that such a potentially dangerous system would ever be allowed to fly," Young insisted: "launch schedule pressure."[17] Challenger, Young feared, had flown in an act of fatalism rather than optimism.

Although providing a rich account of technical deficiency, error, and even possible malfeasance, the Rogers Commission's report concluded that with some remediation of the SRBs and modifications to safety procedures, the shuttle could continue to fly.[18] The commission was not the only external body to examine the disaster: the Committee on Science and Technology in the House of Representatives undertook hearings as well, though the prominence of the Rogers Commission no doubt enabled the Reagan administration to control the interpretation of the disaster more effectively.

Ultimately, the various fact-finding boards reached conclusions with small but significant differences.[19] Technical deficiencies in the SRBs were relatively obvious and easy to pinpoint, but concerns about management processes within NASA were characterized differently by the Rogers Commission and the House investigation. Although generally commending the work of the commission, for example, the House committee took aim at it for suggesting that "poor communication" and "inadequate procedures" on the day of the actual launch had played as large a role in the disaster as the commission claimed. Instead, the House committee concluded that Challenger was felled by "poor technical decision-making over a period of several years," a more damning conclusion that, if it had attracted more notice, might have proven more threatening to the space shuttle program.[20]

The Dissenters

The work of the Rogers Commission was believed by at least some of its members to have been constrained in one fundamental way: whatever its findings, the commission would need to restore public confidence in NASA generally and the space shuttle program in particular.[21] A critical national infrastructure, the space shuttle was too important to be abandoned whatever the findings, and more than anything, the commission would need to validate NASA's decision to press on with it despite the accident. While the commission affixed blame on NASA and its SRB contractor, attention fell mostly on the design of specific components and procedures, not the shuttle's basic architecture, NASA's fundamental competence, or the riskiness of spaceflight generally. This reluctance bred dissent within the commission itself, which soon spilled out into public debates. These varied from indictments of the shuttle's technology and mission, to critical assessments of NASA's management, to issues plaguing all high-technology endeavors, including corporate hierarchies and flawed business communication.[22] The resulting controversy, replete with sociological explanations for failure and competing contestations of responsibility and blame, ultimately became a case study taught in engineering ethics courses about the consequences of poor engineering practice and the mismanagement of risk.[23]

The most influential book on the subject was Diane Vaughan's revisionist 1996 masterwork *The Challenger Launch Decision: Risky Technology, Culture, and Deviance at NASA*.[24] The space agency, in Vaughan's estimation, was infected with a kind of "banality of organization" akin to the philosopher Hannah Arendt's concept of the "banality of evil":[25] NASA and its contractors were hampered in their duties by a bureaucratically dysfunctional work culture exacerbated by "production pressures" and organizational secrecy that prevented engineers from delaying Challenger's launch until conditions became more amenable to safe operation of the vehicle.[26] A sensation when it appeared, *Challenger Launch Decision* remains the foremost work on the disaster and the shuttle program itself.[27] Its lessons have been applied widely to NASA's management generally, and it loomed large over later shuttle controversies.

Vaughan's conclusions expanded upon rather than undermined those of the Rogers Commission, utilizing the evidence that it uncovered (as well as her own interviews), but focusing on the soundness of NASA's management culture rather than the failure of components or the blameworthy actions of particular individuals. While the commission condemned specific technical flaws in the SRBs and deficiencies in analysis and supervision, Vaughan identified instead what she regarded as a kind of pervasive institutional collapse stemming from "an environment of scarcity and competition, an unprecedented, uncertain technology, patterns of routinization, organizational and interorganizational structures, and a complex culture."[28] Long aware that the shuttle did not perform as expected (indeed, the Rogers Commission described the tragedy as a "disaster rooted in history"),[29] NASA and Thiokol created a byzantine and dysfunctional set of launch procedures that unintentionally allowed them to excuse a series of potentially dangerous anomalies in the SRBs: As Vaughan noted, "The attention paid to managers and rule violations after the disaster deflected attention from the compelling fact that, in the years preceding the Challenger launch, engineers and managers together developed a definition of the situation that allowed them to carry on as if nothing was wrong when they continually faced evidence that something was wrong. This is the problem of normalization of deviance."[30]

That NASA managers had accepted this "normalization of deviance" was not the result of conspiracy, cover-up, or other malfeasance, Vaughan argued: NASA and its contractors had not sought to build and fly a defective craft and were not amoral actors risking human life to profit personally or institutionally. Rather, defects in NASA's organizational culture had prevented engineers from accurately assessing and controlling risk. These organizational defects were what doomed Challenger, not, for example, specific acts of misconduct by engineers or politicians that might have created undue pressure to launch in the days leading up to the flight.[31]

Vaughan claimed support for her conclusions in the work of previous scholars, including Charles Perrow's 1984 book *Normal Accidents: Living with High-Risk Technologies*.[32] Three years after the publication of *Challenger Launch Decision*, Perrow released a new edition of his own book mentioning the tragedy. While the discussion of the Challenger accident in the new edition was tantalizingly brief, the placement, on its cover, of the same image of Challenger's destruction found on the cover of Vaughan's book inadvertently made the shuttle *the* symbol of the so-called normal accidents about which Perrow wrote.[33] "Normal accidents," Perrow wrote, are not entirely unexpected: warnings about potential problems usually exist before the failure, but generally only make sense in retrospect. The critical event is often a series of smaller equipment failures, coupled with human error that exacerbates them. Finally, a "negative synergy" combines the manifold small failures into a systemwide problem far worse than the sum of its constituent problems.[34]

In the years that followed the publication of the first edition of *Normal Accidents*, scholars like the sociologist Andrew Hopkins questioned Perrow's model, arguing, for example, that even Perrow's paradigmatic case of the Three Mile Island nuclear disaster did not actually follow Perrow's rubric; instead, negligent managers had discovered glaring problems with critical systems beforehand—and ignored them.[35] Similarly, the O-ring failure and joint rotation issues that destroyed Challenger had been debated from the inception of the program. Challenger's loss stemmed from the relatively simple failure of a single mission-critical system known to work poorly, and human error played almost no role in it. Not surprisingly, given the charge to apply his model to the Challenger disaster in the second edition of *Normal Accidents*, Perrow refrained from doing so. His actual discussion of the accident occupies only two pages into the revised edition, and his conclusions constituted both a deviation from Vaughan's work and from some of his own, earlier ideas. Although placing the accident within the context of systemic failure, for example, Perrow criticized Vaughan for not blaming NASA's management enough for discrete missteps, and for searching instead for a "banality" of management that did not exist.[36] Perrow argued that NASA management had, by 1986, experienced a pathological breakdown due to scheduling pressures: a problem perhaps common to the management of complex technical systems, but outside the norms of competent organizational practice.[37] To Perrow, NASA's deviances were never "normal."

Unconvinced by Vaughan's and Perrow's arguments, a number of other scholars were even less inclined to blame either individuals or NASA management for the tragedy, asserting instead that all large technologies are inherently unstable. Writing two years after the publication of Vaughan's book, Harry Collins and Trevor Pinch, in an essay entitled "The Naked Launch: Assigning Blame for the *Challenger* Explosion," questioned Vaughan's critique of NASA and Thiokol management, calling attention instead to the philosophical challenges that lie beneath any attempt to model physical phenomena in the world, as well as the difficulties associated with interpreting test data of any kind.[38] Drawing heavily upon Vaughan's research, they nonetheless challenged some of Vaughan's major conclusions, noting that while NASA and Thiokol had both been suspicious about the joints between the SRB segments, definitive data were elusive, and even Mason's much maligned instruction for Lund to put on his "management hat" was more likely a statement about the need to weigh conflicting engineering data, not ignore them.[39] That Lund chose to recommend launch was not, the authors suggest, an abrogation of his responsibilities, but an acceptance of them. In Thiokol and NASA, Collins and Pinch found competent if unremarkable management structures, cursed, not with organizational dysfunction, but an almost impossible mandate to make an inherently dangerous technology appear safe.

Vaughan's explanation, Collins and Pinch continued, was compelling not because it was correct, but because, despite its sociological pretensions, it still offered sympathetic heroes and loathsome villains and thus fit easily into contemporary conservative folklore about the suppression of individual autonomy by government bureaucracies:

> The conventional wisdom was this: NASA managers succumbed to production pressures, proceeding with a launch they knew was risky in order to keep on schedule. The Challenger accident is usually presented as a moral lesson. We learn about the banality of evil; how noble aspirations can be undermined by an uncaring bureaucracy. Skimp, save and cut corners, give too much decision-making power to reckless managers and uncaring bureaucrats, ignore the pleas of your best scientists and engineers, and you will be punished.[40]

This narrative, Collins and Pinch argue, was "too simple": the days prior to Challenger's launch were characterized by conflicting opinions, and evidence for the likelihood of a problem with the shuttle's SRB joints was far from overwhelming. Confronted with a potential risk (one of dozens faced every day in the shuttle program), NASA and Thiokol made a reasonable decision that was wrong, not a negligent one. As they noted: "There were long-running disagreements and uncertainties about the joint but the engineering consensus . . . was that it was an acceptable risk." Focusing blame on the engineers and their superiors managing the program, Collins and Pinch argued, turned attention away from the true lessons of Challenger's loss: that spaceflight is inherently risky and "without hindsight to help them the engineers were simply doing the best expert job possible in an uncertain world."[41]

Collins and Pinch's goal was to explore the "golem"—a mindless monster in Hebrew mysticism, made of dirt—at work in modern technologies upon which people tend to have too much confidence. Challenger failed, they suggested, not because engineers were blind to its failings, but because complex machines are fated to fail at some point or another, and that propensity for failure marks a healthy technological research program, not a defective one. As Collins and Pinch noted, an acceptance of unconscionable risk was the norm rather than the exception in spaceflight: early space programs assumed far more risk than the shuttle program did at its height, and many early space programs suffered massive failures, as the constituent technologies underlying liquid-fueled rocketry, at least in its early years, were barely understood. Segmented and reusable SRBs, Collins and Pinch noted, were one of many unknown technologies whose risk had not been conquered.[42]

It is with this assertion that Collins and Pinch risked extending their argument too far, absolving the shuttle's designers of responsibility due to a deficiency in engineering knowledge at the time of the vehicle's creation. In fact, segmented solid rockets were not an unknown technology at the time of the shuttle's development—they were a known hazard, an issue to which the House Committee on Science and Technology referred in its report and about which Rogers Commission members recorded their doubts.[43] (A commission member and the first American woman in space, astronaut Sally Ride, noted during the hearings, "Shuttle astronauts are the only people to ride on solid rockets. Are they safe enough?")[44] Although segmented solid rockets had been used before (particularly to augment the thrust of Air Force Titan III launch vehicles), those employed a different joint mechanism and were not expected to be reusable.[45] The advantage of solid boosters had always been their cheapness and disposability; the reusable, human-rated shuttle SRB's "different, and untried, design" required a development effort that was far more difficult than had first appeared.[46] NASA deputy administrator George Low reminded the administrator, James Fletcher, shortly after the shuttle's authorization in 1972 that the use of segmented solids was so controversial within the Air Force that it likely accounted for its original "neutral recommendation" for the spacecraft's construction.[47]

In fact, Thiokol's initial design for the SRBs was the least technically appealing of the four manufacturer proposals, and a competing bid from Aerojet for a safer, unitary design was rejected by NASA for reasons so unclear that Vaughan dedicated an appendix in *Challenger Launch Decision* to defending the decision.[48] The selection of the Utah-based Thiokol remained clouded by accusations of favoritism by Utah politicians and civil servants like Fletcher, and Vaughan was unable to rule out flaws in the contracting process.[49] While Vaughan struggled to justify NASA's puzzling SRB decision, NASA engineers could not bring themselves to do the same. Engineers like Leon Ray who worked on the SRBs in the mid-to-late 1970s assumed that a complete redesign of the boosters was necessary and inevitable, but "years away." Instead, Ray and his colleagues at NASA and Thiokol satisfied themselves as much as they could by requesting smaller changes that would marginally improve the existing boosters' safety.[50] Contractor proposals for safer liquid-fueled boosters appeared throughout the 1970s and 1980s but did not leave the design stage.[51]

Without such improvements, the side-by-side configuration of the shuttle's orbiter, External Tank (ET), and SRBs ensured that any mechanical problems on launch would threaten the physical integrity of the entire shuttle stack. (This design choice contributed to the losses of both Challenger and Columbia.) The NASA spacecraft designer Max Faget had favored, instead, a single, large, liquid-fueled booster stage, while his colleague, Wernher von Braun, favored launching the shuttle atop the liquid-fueled Saturn V.[52] Faget's vehicle could have sent Challenger into orbit without the need for an ET or SRBs, while von Braun's vertical stack would have shielded the orbiter from damage from booster problems upon launch.[53] Both craft, though, would have been enormous, and extremely expensive. NASA chose segmented solids over more reliable liquid-fueled boosters or unitary solid-fueled boosters principally as a cost accommodation.

Vaughan acknowledged this fact early in *Challenger Launch Decision*, recounting the compromises that led to the TAOS configuration and sympathizing with NASA's need to lobby Congress and the president to approve a shuttle architecture that it did not want.[54] This critical point ultimately factored little in Vaughan's analysis, though, which focused on organizational communication at the time of the disaster rather than the years of missteps that led to it. It may be difficult, thus, to characterize the shuttle's failings as entirely unexpected. This is not the "sloppy management" that Hopkins cited as the true culprit in most "normal accidents," but kind of fatalism bred of bad design.[55] The cause of the Challenger disaster was a design problem that no amount of process work could fix.

Whether the space shuttle's mostly successful early flights convinced its engineers and managers of the craft's safety—a critical element in Vaughan's argument—also remains an open question. She concluded that early flights plagued by only minor anomalies encouraged engineers to move the goalposts on system expectations, lulling engineers into a false sense that the shuttle was safe to fly. Paperwork processes intended to identify problems "rationalized" deviances instead of eliminating them.[56] While efforts to minimize problems and expedite the bureaucratic process that approved the shuttle for flight were commonplace,[57] though, evidence that NASA or

Thiokol engineers were actually convinced by their "normalizations" is less conclusive. Many engineers, in fact, regarded failure as an inevitability and never believed that the multiple layers of flight readiness reviews that seemingly certified the shuttle's airworthiness made the shuttle safe. Rather, one anonymous NASA whistleblower described the reviews as elaborate fictions intended to stroke the ego of aggressive, ambitious Marshall Space Flight Center (MSFC) director William R. Lucas and conceal from him engineers' growing fatalism about the shuttle.[58] As one Thiokol engineer, Boyd Brinton, noted in a Rogers Commission interview, the risk of his car tires blowing out on a highway was the same as that of an O-ring failure at launch, and that Brinton had, in response, merely learned to live with the risk of sudden death every time he rode in his vehicle, just as the astronauts would in theirs.[59] Rather than lulling them into a false sense of the shuttle's safety, the shuttles' successful early flights seemed instead to desensitize engineers to the professional risks that engineers might face if they endorsed the flawed vehicle as flight-ready. Saying "yes" to a launch became easier than saying "no," as a "no" might imperil one's career, but even an ill-fated "yes" would be one of many.[60] This is not the unknowability of technology claimed by Collins and Pinch, but something far more troubling. Indeed, evidence of NASA's recklessness made Perrow reluctant to shoehorn the Challenger disaster into his model of "normal accidents."

None of these analyses, furthermore, suggested that engineers tolerated the shuttle's vulnerabilities because they believed that their presence would accelerate the phased development of the STS that NASA officially rejected[61] but that engineers at all levels virtually assumed would occur once the shuttle demonstrated basic flight capabilities.[62] Rushed into service and pressed to fly more frequently than prudent, the shuttle never truly reached "operational" status: flights required extensive maintenance and even the cannibalization of parts from other orbiters to maintain their readiness to fly.[63] It was these kinds of production pressures that various analysts inside and outside NASA expected would eventually force needed upgrades in the STS.[64] These pressures were features, not bugs: inefficiencies and delays were allowed to fester in the shuttle program out of an expectation that redesign would happen.

With the federal government unwilling to appropriate more money for the STS program, though, funding for these upgrades increasingly appeared likely to come only from launch contracts, so NASA maintained its aggressive flight schedule despite the risks.[65] Planned, systemwide improvements to the shuttle were shelved as greater-than-expected maintenance costs, forcing NASA to focus attention on individual launches rather than large-scale changes.[66] That these more significant changes never occurred, Vaughan conceded, may have been the true cause of the Challenger disaster.[67] But this admission may undermine her other conclusions: the problem with Challenger's "deviances" was not that they were "normal," it was that they were not "deviances."

Ultimately, scarcely any of the arguments made about Challenger's demise in the years after the Rogers Commission deliberations were entirely novel: disagreements as to the cause of the tragedy had extended back to the commission itself. Its conclusions had not been unanimous among members of the group, and the consensus report concealed differences of opinion regarding the safety of the STS. Richard Feynman found that the commission appeared to have little authority to evaluate the safety of the shuttle itself, only to ascertain the specific causes of one accident. When Feynman began his own investigation of the shuttle's safety, the commission's leadership marginalized him, and Feynman eventually threatened to withdraw his endorsement from the final report unless it tempered its enthusiastic evaluations of the STS. Unconvinced by evidence of NASA's supposed management failings, Feynman instead focused on fundamental problems with the shuttle's design, exploiting information provided to him by Ride, Air Force major general Donald J. Kutyna, and mostly low-level engineers within NASA.[68]

If NASA management was to blame at all, Feynman explained in a dissent appended to the commission's report, it was only for its unreasonable public claims about the safety of the space shuttle. Feynman's own analyses, and those of NASA engineers he consulted, indicated a failure rate for the shuttle of approximately 1-in-100 flights, quite close to the actual historical figure. Senior NASA management, though, had promised publicly that the shuttle would fail only once in 100,000 flights, a wild overestimate unsupported by mathematics or engineering practice.[69] A 1-in-100 chance of death with each flight represents a kind of risk with which combat and test pilots have long been familiar, Feynman asserted, but few members of the public would knowingly travel in a passenger aircraft with such low reliability. McAuliffe's interviews prior to STS-51L confirm her misapprehensions about risk: she dismissed the idea that flying the shuttle was a "'dangerous thing to do'" and likened it to commercial airline travel (with its 1-in-750,000 failure rate during the 1980s).[70] NASA's public exaggerations about the shuttle's safety, and the Rogers Commission's endorsement of the shuttle despite its obvious weaknesses, Feynman concluded, were hoaxes necessary to sustain the myth that the shuttle was safe enough for the general public to fly:

> In any event this has had very unfortunate consequences, the most serious of which is to encourage ordinary citizens to fly in such a dangerous machine, as if it had attained the safety of an ordinary airliner. The astronauts, like test pilots, should know their risks, and we honor them for their courage. Who can doubt that McAuliffe was equally a person of great courage, who was closer to an awareness of the true risk than NASA management would have us believe?[71]

Collins and Pinch later reiterated this point, noting the discrepancy between the shuttle's imagined safety and the danger that it presented to its crews:

> There is a lesson for NASA here. Historically it has chosen to shroud its space vehicle in a blanket of certainty. Why not reveal some of the spots and pimples, scars and wrinkles of untidy golem engineering? Maybe the public would come to value the shuttle as an extraordinary human achievement and also learn something about the inherent riskiness of all such ventures. Space exploration is thrilling enough without engineering mythologies.[72]

Such candor about the space shuttle no doubt would have had a salutary effect on the program. It might have focused attention on the actual sources of risk, which were not merely the limits of theoretical knowledge about the physical world or the pathologies of large organizations. Instead, engineers were tasked with producing and operating a spacecraft that was actually less robust than they wished it to be.

Ultimately, an explanation of technological failure rooted in work processes alone is neither compelling nor useful. Vaughan may well have been correct about defects in NASA's safety culture, but "normalization of deviance" was most likely a symptom rather than the cause of the shuttle's difficulties. Indeed, Young's and Feynman's evidence undermines Vaughan's assertion that NASA's broken safety culture limited knowledge about defects and created a false sense of security on the part of NASA's engineers. The agency had created procedures to make acceptance of the shuttle's flaws easier to manage politically, but these supposed fixes were never accompanied by actual confidence in the safety of the system. And whether engineers and managers were amoral or negligent or had merely lulled themselves into complacency over the shuttle's underlying technical flaws may be less important than the fact that, without the funds to move beyond the STS, its flaws persisted, even after being publicly exposed and scrutinized.

The New Normal

Given the shuttle's defects, it is tempting to imagine the loss of Challenger as the wake-up call that should have ended the STS. The history of the American space program, though, offered little precedent for such a radical change of course. When three astronauts perished in a launch-pad fire during the test of the Apollo AS-204 spacecraft in 1967, investigators attributed the tragedy to faulty engineering and poor management but resisted calls to suspend the Moon Race. To do otherwise was not politically palatable, and Project Apollo continued after a hiatus, as indeed, did the STS after the Challenger disaster. While the Space Race had become a more amorphous affair in the 1980s, the political will to end American human spaceflight was similarly absent. Instead, NASA attempted what fixes it could on a reasonable budget and attempted to revive the shuttle while turning increasingly to the question of postshuttle space infrastructures (described in chapter 6). Neither problem proved easy to solve.

If the purpose of the Rogers Commission was to restore public confidence in the shuttle, its mission succeeded only partially. In the wake of Challenger's loss, NASA suspended shuttle flights for over two years, but the modifications to shuttle hardware made in response to the accident were relatively minor. To ameliorate launch risks, NASA and Thiokol redesigned the SRB joints,[73] and NASA once again provided pressure suits for crew members to wear upon launch and reentry, though the value of this precaution had always been questionable. (This change was one of several that created a kind of "safety theater" around the space shuttle: routine performances of contingency procedures intended to create confidence in an unreliable technology.)[74] These measures did not reduce the STS's underlying vulnerability; instead, NASA restricted mission objectives and the composition of crews, attempting to avoid the public relations debacle that came with suggestions that the shuttle was an appropriate craft for members of the general public to fly. Replacement of the SRBs with liquid-fueled boosters was repeatedly studied and dismissed despite the fact, as Max Faget noted at the time, that the SRBs accounted for 57 percent of the shuttle's ascent risk.[75]

Most critically, President Reagan made an about-face and cleared the shuttle's manifests of the payloads that it would have carried for private industry. NASA's entrance into the private satellite launch market was always fraught: the move placed the government in competition with US commercial rocket-makers and monopolized the business that they relied upon to remain profitable. More expensive to operate than competing foreign space launch systems, the shuttle drove commercial launch business away from the US even before Challenger made the shuttle appear dangerously unsafe.[76] The Air Force, whose skepticism about the shuttle was present well before its initial authorization in 1972 and continued long after,[77] mostly ignored the commission's more optimistic findings about the overall reliability of the shuttle and withdrew military payloads from the STS. While a few previously scheduled payloads eventually flew, the Air Force and National Reconnaissance Office were largely unmoved by the Rogers Commission's assurances that the shuttle could remain a reliable military workhorse. Having never completely abandoned the use of expendable launch vehicles, the Air Force greatly expanded these efforts after Challenger, shifting payloads to Titan IV launch vehicles (which suffered SRB failures of their own), eventually authorizing a program to create cheaper launch vehicles with improved performance.

With a few strokes of a pen, the economic rationales that had undergirded the shuttle's development were eviscerated, and the space shuttle, in the years that followed the Challenger disaster, began to operate as previous American spacecraft had—as experimental scientific vehicles not expected to turn a profit. Although plans to fly private citizens aboard the space shuttle had never been more than a token effort, NASA's overtures toward democratizing the space program ceased immediately. Almost overnight, expectations that NASA's shuttles would carry members of the general public—or that private industry might fly fleets of vehicles packed with tourists[78]—disappeared. Other missions deemed to hazardous to risk a shuttle crew also fell by the wayside. Yet while voices calling for the retirement of the shuttle became louder following Challenger, the endorsement the craft received from the Rogers Commission ensured its continued operation despite the collapse of its economics.

With expectations for its performance greatly diminished, post-Challenger restrictions on the shuttle's use gave the STS program sufficient breathing room (including smaller payload manifests and reduced launch frequency) sufficient to enable the shuttle to succeed as an experimental orbital launch vehicle comparable in cost and risk to previous human spacecraft programs. The space shuttle would not save money, but its operation could continue within NASA's current budget, once again raising the specter of the sunk cost fallacy. A backlog of unlaunched payloads, for which NASA otherwise would need to purchase billions of dollars' worth of expendable launch vehicles, helped ensure that the shuttle would fly until the orbiters wore out or suffered another catastrophic failure, at which time commentators expected that NASA would retire the STS for good.

Although the loss of Challenger may have punctured the space shuttle's reason for its existence (as well as NASA's decades-old "aura of competence"),[79] it was only after the tragedy that STS flights actually became routine; indeed, shuttle flights before 1986 constituted only a fraction of missions that the orbiters flew in their thirty-year service life. The loss of Challenger may have brought the era of NASA's space truck to a close, but it also led to mostly successful reconfiguration of the STS as an expensive, maintenance-intensive, heavy-lift launch vehicle, a transformation that led to the shuttle's most productive years in service, especially in its orbiting of large scientific satellites.[80] These achievements, though, brought the same mix of success and disappointment as previous space activities: the shuttle could not, ultimately, restructure humanity's relationship to space.

As a tool of laboratory science, the space shuttle continued to be an imperfect vehicle after Challenger, though it offered capabilities not found in previous spacecraft. Routine flights by mission specialists resumed, and the pressurized Spacelab and the subsequent, similar SPACEHAB modules greatly increased the internal volume of shuttle crew spaces and enabled the carriage of automated experiments for microgravity research, astronomy, and other scientific fields. (By 2007, the number of life sciences experiments flown had exceeded 2000.) Yet unlike with research on the ground, shuttle experiments were limited to the craft's typical fourteen-day mission length and had to be prepared well in advance of flight. Damage or errors that might result in an experiment being repeated on Earth brought complete failure in space. Constrained by an externally driven launch schedule planned years in advance, even successful experiments were seldom repeated (despite the low cost involved) in order to make room for other payloads. Despite its multibillion-dollar cost, Spacelab's pressurized modules flew relatively infrequently. And while the experiments routinely carried aboard shuttle flights generated a large quantity of scientific data, shuttle operations remained expensive through the end of the program. Each flight cost an estimated $500 million (comparable to the cost of Apollo-era launches), and payload rates into low-Earth orbit hovered at approximately $10,000 per pound, more than ten times what the program's most optimistic forecasts had predicted and comparable to the rate achieved by expendable launch vehicles.

NASA continued, through the 1990s, to promote the unique capabilities that the shuttle offered, but it was not always clear what these capabilities were. Foremost among the shuttle's ambiguous achievements in this period was the launch of the Hubble Space Telescope (HST) in 1990 and its subsequent retrieval, repair, and servicing during five additional flights from 1993 to 2009. Although this demonstrated the shuttle's ability to maneuver to a precise location in orbit, retrieve a satellite, repair it, and release it intact, doing so proved wildly expensive—less an argument for the shuttle's utility than a demonstration of a capability for its own sake.

Among the oldest concepts for an orbiting scientific platform, the craft that became the HST was the subject of aggressive lobbying from the start of the Space Age; by positioning telescope optics outside Earth's atmosphere, astronomers hoped to achieve multispectral imagery of unprecedented sensitivity and clarity. This information would enable the study of distant and dim deep-sky objects, and the study of light from the deep past, which would provide information on the origins of the Universe. Including a range of cameras in the vehicle's design (including a planetary camera that some thought useful only for public relations purposes) also enabled spectacular images of planets, nebulae, and other well-known interstellar objects.[81]

Nearly identical to the KH-11 photographic reconnaissance satellite, the HST fit snugly into the space shuttle's cargo bay: indeed, KH-11 or its predecessor, the KH-9 "Big Bird," were likely the payloads that US deputy secretary of defense Dave Packard feared were driving the shuttle's configuration in 1971. Developing alongside the space shuttle, the HST was designed to be launched and serviced by it instead being merely replaced when defunct, a controversial decision that enabled flexibility and longer operating life but tethered the satellite's future to continued shuttle operations, an issue that proved critical as the shuttle program wound down.[82] Like the military reconnaissance satellite on which it was based, the HST experienced significant problems during development, including schedule delays and cost overruns with prime contractor Lockheed and its subcontractors. Indeed, the discovery, shortly after the HST's deployment from the Discovery shuttle on STS-31, that the primary telescope mirror of the satellite had been ground incorrectly by its manufacturer, Perkin-Elmer, provided yet another public relations fiasco for NASA, in which the agency was blamed for the sloppy oversight of an expensive space project.

Like the Challenger disaster, the HST's vision problem was a single failure with serious consequences. Unlike with Challenger, the mirror's flaws did not compromise the vehicle's structural integrity or safety. Later shuttle flights not only enabled the installation of corrective optics to compensate for the flaw in the mirror's fabrication, but also upgraded or replaced key sensory and navigational equipment, extending both the operational life of the telescope and its capabilities. During the wind-down of the space shuttle program following the loss of Columbia in 2003, the continuing need to visit the HST for repairs to extend the telescope's life created significant controversy: rejecting a mission to the HST in 2004, the cost-cutting NASA administrator Sean O'Keefe faced pressure from NASA's scientific community, politicians, and the press.[83] His replacement, Michael Griffin, approved the flight, which occurred in 2008. While hailed at the time as a necessary risk to preserve a valuable scientific asset, the episode demonstrated both the potential and limitations of human-serviced scientific satellites. Turning HST's failure into a justification for the shuttle's continued operation, NASA and its supporters argued that no craft other than the shuttle orbiters could have successfully captured the HST in orbit and conveyed both the equipment and personnel necessary to repair it. Critics, though, were quick to point out that many HSTs could have been funded with the shuttle's budget.[84] Indeed, expendable Air Force launch vehicles throughout this period maintained a fleet of KH-11 satellites without the need for repair by astronauts, by simply replacing satellites that were no longer serviceable.

During the 1990s, shuttle flights launched many scientific satellites, but these efforts seldom galvanized significant public enthusiasm and could have been launched on expendable vehicles at the same cost. They included the Compton Gamma Ray (1991) and the Chandra X-Ray (1999) observatories into Earth orbit, as well as a number of deep-space probes. Magellan, launched in 1989, mapped the surface of Venus before undertaking a controlled entry into the Venusian atmosphere. Galileo, launched the same year, became the first spacecraft to orbit Jupiter, and it dispatched a probe into the planet's dense cloud layers. (Original plans to launch it aboard the shuttle had given rise to the "Death Star" jokes.) Ulysses, launched in 1990, utilized a Jupiter-fly-by to catapult itself into solar orbit out-of-plane with the Solar System, enabling examination of the Sun from all latitudes. Regular shuttle flights during this period also enabled the launch and recovery of spacecraft intended to provide data about long-term exposure to the space environment. Challenger's deployment of the Long Duration Exposure Facility in 1984 and Columbia's retrieval of the satellite in 1990 provided unique information regarding the orbital debris and micrometeoroid environment of low-Earth orbit, as well as over fifty additional experiments. Further experiments in 1996 involved the launch and release of the Orbiting and Retrievable Far and Extreme Ultraviolet Spectrometer-Shuttle Pallet Satellite II (ORFEUS-SPAS II) pallet, which contained ultraviolet spectrometers to study stars and the interstellar environment.

Within a decade after Challenger's loss, though, flights by the three remaining orbiters (Columbia, Discovery, and Atlantis, eventually supplemented by Endeavour) were not only routinely carrying Spacelab and SPACEHAB modules, but also deploying and retrieving satellites and visiting and servicing a space station in a manner close to what its designers had envisioned. By the late 1990s, the shuttle had once again established something close to a regular launch schedule, although NASA had effectively been abandoned the major impetus for its construction—cost savings. In its place, the shuttle became a peculiar space infrastructure that, while expensive, appeared to satisfy two significant needs simultaneously: the continuance of American human spaceflight and the launch of NASA's heavy scientific satellites: the latter was not an insubstantial role, but a crewed spacecraft was not required to achieve it. Neither did the shuttle offer a particularly efficient solution to either problem: the shuttle was larger than necessary to merely carry astronauts, lacked suitable abort modes, and required too much hands-on piloting. As a cargo carrier, it had unnecessary human stowaways whose physiological needs complicated launches and used up valuable mass and volume allowances. Instead of splitting the space program into separate vehicles for passengers and cargo, NASA managed to bootstrap its floundering human spaceflight program by piggybacking additional crew members onto satellite launches that would otherwise have been unpiloted. And while it had done much to ameliorate the vulnerability that felled Challenger, the STS remained a dangerous technology with many more likely modes of failure.

The Columbia Tragedy

The destruction of Columbia on reentry during mission STS-107 on February 1, 2003, like the earlier loss of Challenger, occurred as the result of a

Figure 5.3
The STS-107 crew members pose two years before their ill-fated 2003 Columbia flight. From left: Mission Specialist David Brown, Commander Rick Husband, Mission Specialist Laurel Clark, Mission Specialist Kalpana Chawla, Mission Specialist Michael Anderson, Pilot William McCool, and Payload Specialist Ilan Ramon. NASA photo.

problem that was too dangerous to ignore, yet too difficult to solve (figure 5.3). As with Challenger, the failure occurred upon launch and varied from many previous, similar failures only in degree.[85] Again, the problem stemmed directly from the choice of booster for the vehicle: an expendable, external propellant tank and SRBs strapped alongside the orbiter rather than below it. Instead of the SRB failing, though, Columbia was struck by an insulating foam shed from the ET, which damaged the orbiter so much upon launch that it could not safely return to Earth.

The $30 million ET that held the liquid hydrogen and liquid oxygen for the SSMEs was the simplest element of the STS, and the only part that was fully disposable; the lack of reusability of the ET, though, did not mean that its operation was trouble-free. In fact, it had been a constant source of concern throughout the program. An aluminum tube 154 feet tall and 28 feet wide, the tank contained two pressurized compartments: a large bottom tank for hydrogen (chilled and compressed into a liquid) and a smaller top tank for liquid oxygen, separated by a mechanical intersection. Large pipes conveyed the contents of the tanks to the shuttle orbiter, whose SSMEs expended the fuel and oxidizer during its ascent. Covering the tank was a spray-on foam insulation often described as "orange," "brown," or "rust"; on the earliest shuttle flights, the insulation was painted white to protect it from ultraviolet radiation (and match the rest of the shuttle stack). Jettisoned just before the shuttle reached orbit, the ET coasted on a suborbital trajectory before reentering the atmosphere over an ocean area clear of naval traffic. The shuttle orbiter's own Orbital Maneuvering System (OMS) engines in pods attached to the aft fuselage then provided the last amount of thrust necessary to circularize the shuttle's orbit and render it stable.

To prepare for launch, shuttle engineers would mount the shuttle onto its launch platform, move it to the launch site, and then fill its fuel tanks. After fueling, the ET's cryogenic propellants would slowly begin to boil, and the insulation kept heat out as much as possible before launch, preserving sufficient liquified propellants for the launch. Ambient water vapor on humid Florida mornings routinely condensed on the exterior of the ET and froze, hardening the insulation and making it prone to cracking on liftoff, especially in the areas near supports that attached the tank to the orbiter (called the "bipod ramp"). Loss of ice, foam, and other debris from the tank during ascent had been problems on many previous missions; flecks of white paint from the ET had struck Columbia's windows on STS-1, and orbiters on later mission frequently returned from flight with gouges in its thermal protection system (TPS) caused by shedding paint, ice, or foam, some of which was large enough to breach the TPS entirely and expose the fuselage.

Until February 2003's STS-107 flight, the shuttle's TPS had never been so compromised as to threaten the survival of a crew: failures at that point had degraded exterior surfaces on the craft but were prevented from causing further damage due to the random location of the breaches on top of aluminum panels and other protective structures.[86] On Columbia's final flight, however, a large chunk of icy foam peeled off the nose of the tank and struck the left wing's leading edge, where brittle carbon tiles protected it from the worst frictional heating of atmospheric reentry, and where hollow wing structures beneath the panels offered little further protection (figure 5.4). (After the accident, speculation ensued as to whether NASA's suspension of its practice

Figure 5.4
The space shuttle Columbia lifts off from KSC, Florida, on the morning of January 16, 2003. The left wing's leading edge, in gray, would soon be pelted with foam from the orange ET above it. NASA photo.

of painting the orange foam covering the ET white before earlier shuttle flights had made that shedding of foam more likely.)

Protection of the shuttle's delicate wings and fuselage during reentry was both one of the earliest technical challenges for the program and one of the most enduring. Because the shuttle, like all previous human spacecraft, carried insufficient fuel to slow its orbital motion completely, it relied upon a short firing of its OMS engines to slow it and cause it to descend, whereupon atmospheric drag would complete its deceleration. This friction produces tremendous heat, and space vehicles use a variety of technologies to manage it. NASA's Project Mercury spacecraft had employed a metallic heat sink covering the portions of the vehicle subjected to the greatest reentry heating, and Project Gemini and Apollo vehicles employed ablative shields composed of an aluminum honeycomb packed with a phenolic resin. The shuttle employed a variety of novel reusable heat-resistant materials designed *not* to char or degrade during heating, including silica tiles and a small number of gray, reinforced-carbon-carbon (RCC) panels on portions of the craft exposed to the greatest heating. Although appearing to be single, molded, solid sheets, the panels were actually a composite material composed of layers of graphite-impregnated rayon and phenolic resin baked in an oven and sealed against cracks.[87] This technology had the advantages of low mass and reusability, but it traded these qualities for physical durability. The RCC panels in the nose cone and on the leading edges of the wings were known to be extremely brittle, even though they covered forward-facing portions of the orbiter that were likely to be pelted with launch debris. Small punctures in the TPS were not generally a concern; unusual about the debris that struck STS-107 was its relatively large size, as well as the peculiar misfortune of an impact in one of the areas in which reentry heating was the greatest.

Once the foam separated from the bipod ramp, its high air resistance relative to its mass rapidly decelerated it, until the difference in speed between the foam chunk and shuttle was 545 miles per hour: this was the speed with which the foam struck RCC panel 8 of Columbia's left wing.[88] NASA become aware of the impact of foam on the day after the launch after studying film footage of it, but its response to the potential problem was less urgent than it had been even for previous, lesser occurrences of the same issue. Noted the Columbia Accident Investigation Board (CAIB), "When a debris strike was discovered on Flight Day Two of STS-107, Shuttle Program management declined to have the crew inspect the Orbiter for damage, declined to request on-orbit imaging, and ultimately discounted the possibility of a burn-through."[89] NASA ground experiments suggested that no impact of frozen foam would threaten the orbiter's physical integrity, and NASA management refused an offer by the National Reconnaissance Office to examine the orbiter using spy satellites (as had been done with Columbia's first launch in 1981), or permit Columbia's crew to undertake extravehicular activity (EVA) to inspect or repair the damage.

There is no doubt that upon reentry, the hole in the left wing's leading edge proved the craft's undoing. Streaking through the upper atmosphere at twenty-five times the speed of sound, Columbia's most capable heat shield technology gave way: superheated gas ("probably exceeding 5,000 degrees Fahrenheit")[90] entered the hole in the wing and destroyed its aluminum framework. Alarms sounded in Columbia's cockpit as temperature sensors in the wing and landing gear began to redline and then fail, suggesting the imminent loss of the wing itself.[91] The crew had little time to react and no real options: bailout from the shuttle while it was in hypersonic flight and bathed in plasma produced by frictional heating was impossible, and the vehicle did not carry sufficient fuel for its smaller OMS engines to return to orbit. The probable tearing of the left wing from Columbia seconds into the failure doomed the craft: thruster firings that attempted to stabilize the craft would not be able to overcome the aerodynamic forces produced by such unbalanced pressure loads. Fatally unstable, Columbia tumbled and broke up from aerodynamic forces and extreme heating, killing the crew and spreading wreckage across Texas.[92]

The fact that the failure occurred over land enabled NASA to collect and reassemble the pieces of the vehicle that survived reentry and landing. The investigations that followed the tragedy (particularly the CAIB, created by NASA and led by retired admiral Harold W. German, Jr.) followed the pattern of the earlier Rogers Committee with almost shocking similarity: again, evidence revealed a persistent design flaw that produced numerous anomalies in flight, but that were ignored and minimized to keep the shuttle flying. Unlike the earlier investigation, though, the work of the CAIB was informed by nearly two decades of scholarship on shuttle management failures, and Vaughan's assessment of the Challenger disaster became central to analyses of the later tragedy. In *Challenger Launch Decision*, Vaughan had described how a frustrated and overwhelmed management structure within NASA during the mid-1980s gradually came to ignore glaring risks associated with the shuttle and convinced a new generation of space engineers to accept aberrant performance as acceptable if it did not result in death.[93] Reviewing the accident with the benefit of Vaughan's analysis before it, the CAIB noted that, as in the case of Challenger, unknowns present in the STS had lingered amid an insufficiency of reliable engineering test data. Without good data to demonstrate safety, "shuttle managers used past success as a justification for future flights."[94] Again, wishful thinking replaced careful analysis.

While the Rogers Commission was somewhat less inclined to focus on problems in organizational dynamics and business communication, the CAIB embraced these concerns. Singled out for particular blame were modern styles of technical briefings by engineers, particularly the replacement of detailed reports with digital presentation slides, mostly prepared with Microsoft's market-leading PowerPoint software.[95] The CAIB cited these slides as at least a contributing cause in NASA's failure to properly appreciate the danger that Columbia faced from detached foam.[96] Edward Tufte, the leading critic of this method of discourse, later wrote about PowerPoint's propensity to obscure truth and diminish important findings with generic slides of list-based data.[97]

The CAIB's conclusions appeared to satisfy political leaders and the public, in part because they, like those of the Rogers Commission, criticized obvious technical flaws in the shuttle, as well as persistent failures of the shuttle's management, without questioning the value or purpose of the shuttle itself. While the CAIB noted that the space shuttle had not met its designers' hopes as an operational vehicle, they argued that as an experimental craft, it remained a critical part of the national technical infrastructure:

> Although an engineering marvel that enables a wide-variety of on-orbit operations, including the assembly of the International Space Station, the Shuttle has few of the mission capabilities that NASA originally promised. It cannot be launched on demand, does not recoup its costs, no longer carries national security payloads, and is not cost-effective enough, nor allowed by law, to carry commercial satellites. Despite efforts to improve its safety, the Shuttle remains a complex and risky system that remains central to U.S. ambitions in space. *Columbia*'s failure to return home is a harsh reminder that the Space Shuttle is a developmental vehicle that operates not in routine flight but in the realm of dangerous exploration.[98]

Noting the physical vulnerabilities that felled Columbia, the CAIB insisted upon minor improvements to the shuttle's foam insulation and more and better photography and inspection capabilities on launch and in orbit, but it refused to recommend the cessation of the program. When the subsequent flight of Discovery in 2005 revealed similar problems with shedding foam,[99] NASA suspended shuttle flights until 2006.

On close examination, Vaughan's "normalization of deviance" and Charles Perrow's model of "normal accidents" explain the loss of Columbia no better than they did that of Challenger. Not only had the inherent risks of the shuttle remained long after Challenger's loss, but the vulnerabilities that destroyed the second orbiter were, like those that destroyed the first, earlier accommodations to cost that knowingly increased risk. The shuttle's TPS differed markedly from the metal heat sinks used for NASA's Project Mercury and the ablators used for Projects Gemini and Apollo: these were physically robust and relatively invulnerable to physical damage. Heat sinks were, furthermore, reusable, while ablators could be reapplied to space vehicles after their return to Earth. Neither technology, though, suited a winged space vehicle or one intended for rapid and inexpensive reuse. Metal heat sinks were extremely heavy, while ablative coatings required careful reapplication when used on previous craft like the X-15. Some heat-resistant metals showed promise as replacements for the shuttle's skin, but the Air Force had barred the shuttle from employing titanium construction, and even metal heat shields required delicate coating to prevent oxidation. A physically robust TPS that offered maintenance-free reuse simply did not exist at the price point and under the manufacturing restrictions set by the Air Force, so designers instead chose the only technology available that would come close.

While the shuttle's TPS had met its minimum requirements, the damage that it incurred on most flights was less a normalized deviance than a source of constant worry on the part of engineers and managers, from the shop floor to the managers at NASA headquarters. It was not that NASA had ignored the danger that destroyed Columbia; rather, the risk just paled in comparison to other equal or greater dangers, about which NASA could not do anything without redesigning the shuttle. NASA appeared to operate with two sets of risk calculations: an optimistic public assessment and an internal one characterized by deep misgivings that it failed to share with the president, Congress, or the public. NASA had, as with Challenger, cultivated a public-facing belief that the shuttle was safer than it actually was, even flying nonprofessional crew members as Payload Specialists and engaging in public relations flights despite the known dangers of the system. As in the case of Challenger, analysis subsequent to the Columbia disaster noted that NASA's culture of safety was always illusory, and the agency's mandate to explore space made its work inherently dangerous.[100]

Not surprisingly, Columbia's loss was also a poor candidate for Perrow's "normal accident" analysis, on which Vaughan had based her own study. Like the SRB joints, the vulnerability of the TPS was an early and persistent subject of engineering concerns, dating back to the troubled installation of heat-resistant tiles on Columbia in 1979 and the loss of several in flight in 1981. On a semisecret 1988 Atlantis mission to deploy a spy satellite, NASA's Mission Control Center (MCC) responded to reports by the crew of significant launch damage to the TPS with disbelief and dismissal:[101] damage as extensive as that which the crew described could not be repaired in flight and hence was best ignored. Recalled Mission Specialist Mike Mullane:

> *Say what?!* We couldn't believe what we were hearing. MCC was blowing us off. There was no discussion of having ground telescopes take some photos of Atlantis to possibly get a better view of the damage. There was no discussion of having us power-down the vehicle to give us maximum orbit time to deal with the problem. There was no indication whatsoever that MCC thought we had a serious problem.[102]

NASA's response to the crew's concerns, Commander "Hoot" Gibson recalled in a 2009 article in *Spaceflight Now*, was galling: "NASA does amazing things when they've got their back against the wall," he said. "Like Apollo 13. I've seen us work out some really dramatic things in some of the missions when we had on-orbit problems and we did in-flight maintenance and things like that. You never know what you could have done because you didn't try."[103]

The 2003 Columbia accident was the result, not of the cascading, unanticipated failure of multiple mechanical systems, but of a basic structural component, known to be unreliable, that broke in flight, damaging another vulnerable component in its path in a manner that had happened on many previous occasions. NASA had also previously debated replacing the RCC panels on the wing edge with a hardier material but had failed to do so for cost reasons. Once again, random chance played a greater role in the problem than human error: while debate raged about whether Challenger should have launched in cold weather, Columbia's final launch was relatively uncontroversial. Furthermore, no "negative synergy"[104] was required to fell an orbiter with a hole in its wing: this kind of failure was a worst-case scenario for which there was no known solution or means of rescue. Destruction of a wing and failure of the TPS were unrecoverable problems for the STS, and indeed were from the inception of the program.

The failure that had destroyed Challenger offered NASA's astronauts and ground controllers no time to assess the problem and recommend solutions; the damage that Columbia incurred upon liftoff would not create grave hazards until reentry, giving NASA weeks, potentially, to diagnosis the problem and take corrective measures. That NASA did not undertake these efforts offers a point of difference between the two cases that creates new difficulties in applying Vaughan's and Perrow's analyses to the loss of Columbia. Once Columbia's wing was damaged, the shuttle crew had thirty days to manage the contingency before the crew consumed the orbiter's supplies, but the vehicle offered no effective on-orbit rescue technology (another accommodation to cost). An Apollo-style escape pod mounted in the shuttle's cargo bay was deleted early in the shuttle's design process for cost and mass reasons. A lack of handholds on the exterior of the craft made it challenging for spacewalking astronauts to repair the wing, and they had no proper materials on board with which to do so anyway. NASA had no second orbiter on standby with which to rescue the crew, and an insufficient number of spacesuits to protect them even if one had been. Subsequent analyses, though, questioned the severity of these obstacles,[105] suggesting that the insertion of virtually anything into the hole might have preserved the orbiter's aerodynamics, and in any event, the astronauts deserved some kind of effort to save their lives, even if it were unlikely to succeed. CAIB investigators agreed, concluding that NASA could have attempted all these options, as they posed no more danger to the crew than that which they already faced.[106]

Instead, in a jarring act of fatalism, NASA headquarters under O'Keefe determined that any damage to Columbia that was serious enough to compromise its structure likely could not be repaired anyway, and thus NASA would make no attempt to repair the damage. O'Keefe, an accountant and self-described "bean counter," had been appointed by President George W. Bush to reduce NASA's expenditures, and critics later seized upon his tenure as the source of at least some of the agency's woes during the period, including the flawed, almost parsimonious management of the Columbia crisis.[107] Indeed, on 2005's STS 114 Discovery flight, the first "return-to-flight" mission of the space shuttle after the Columbia disaster, as the longtime shuttle program manager Wayne Hale recounted, a large piece of foam again separated from the ET and struck the orbiter's wing, virtually re-creating the accident that felled Columbia. Although better in-flight inspection protocols ruled out any significant damage to the craft, NASA had not, in the intervening two years, been able to ameliorate the vulnerability.[108]

Ample public debate about the significance and potential danger of Columbia's problem followed NASA's discovery of it. What resulted was less a misunderstanding of risk than an almost incomprehensible neglect of a known problem: a lethargic, fatalistic response driven almost entirely by expected value calculations. Organizational dynamics played less of a role in the tragedy than bad choices made for bad reasons, in some cases by people lacking the expertise to make them. While none of the rescue techniques suggested by the CAIB may have saved Columbia, the fact that none were attempted, over the objections of NASA's astronauts and engineers, seemingly marked a new and more dangerous form of management failure than what Vaughan had described.

Conclusion

In the wake of the Challenger disaster, mounting problems with the shuttle prior to the launch gave the previous year's hectic schedule and slow accretion of problems the appearance of a failing program beset by management and maintenance issues. These seeming failures were worsened by NASA's constant overestimation of the shuttle's capabilities, which created political problems for the craft even though the shuttle was about as reliable as any previous human spaceflight program at that point. While the management of the shuttle program was imperfect, problems with the shuttle stemmed mainly from the limitations of an architecture chosen as a compromise between safety and cost. A focus by later scholars on organizational dynamics had the unintentional effect of making the disaster appear to be the inevitable product of a dysfunctional work culture rather than the result of self-interested and callous decisions made by particular individuals decades earlier.

Rather than ending the space shuttle program, the loss of Challenger in 1986 transformed it from an operational vehicle with a routine mission to an experimental craft bearing the history of past catastrophe. For NASA, newfound awareness of the shuttle's vulnerabilities prompted a reassessment of the program's goals and created a more limited vision of its transformative potential. A commitment to maintaining America's only human spacecraft led NASA to imagine new uses for the craft and to embark on a program of scientific exploration, in part to demonstrate the shuttle's continued utility. In the fifteen years of shuttle operations following the loss of Challenger, NASA found a new identify for its spaceplane, though one not quite as lofty as it had once imagined. Even after NASA adjusted its work processes and fixed certain glaring flaws in the SRBs, the STS continued to underperform, requiring more maintenance and offering higher costs and levels of risks than promised. While the STS continued to fly after Challenger, it did so as an experimental craft flying infrequently with restricted payloads.

Signs of the shuttle's continuing vulnerability into the late 1990s and early 2000s were readily apparent despite many successful launches. Expectations of airplanelike reusability had long since disappeared, with shuttles undergoing costly overhauls and partial disassembly between flights to ensure the reliability of the major components, especially the shuttle's main engines. These still absorbed enormous pressures and temperatures in flight, and the shuttle routinely returned from space with its TPS dinged by innumerable holes and gauges from impact by unknown debris upon launch. While the STS continued to fly, its value derived almost entirely from its capabilities as a heavy-lift vehicle and its unique political status as the only extant American human spacecraft. The frequency with which the shuttle flew during this period and the ambitiousness of its missions despite its inherent vulnerabilities suggests a kind of cognitive dissonance later noted by the CAIB.

Too expensive to be abandoned, NASA's space shuttle succeeded in becoming an essential transportation infrastructure: the shuttle had fallen short of its original goal of routine spaceflight and low-cost operation, but its large payload bay and the mere fact of its continued operation prolonged the use of the STS for another quarter-century after the Challenger disaster and nearly a decade after Columbia. As America's only human spacecraft, and one of the few large enough to lift heavy payloads into space, it became a finicky but essential piece of the national technical infrastructure—instead of serving as the blueprint for a cosmic future, the shuttle became among the first of a series of institutions "too big to fail."[109] So long as NASA flew the shuttle, less money would be available to fund alternatives,[110] but lacking any alternatives, the shuttle still needed to fly.

On January 14, 2004, following the loss of Columbia, President Bush announced the termination of the space shuttle program in connection with a new series of space exploration goals, but this decision was the beginning of an elaborate demobilization rather than an abrupt end.[111] Streaking into the skies for seven more years, the space shuttle existed as a working monument to bad design, necessary only because NASA possessed no vehicle capable of taking its place. Improbably surviving these crippling setbacks, the shuttle continued to operate under the growing weight of its history.

6 The Quest for Alternatives, 1972–2011

In films about the space shuttle during its early years—like the 1983 television movie *Starflight: The Plane That Couldn't Land,*[1] and 1986's Hollywood flop *SpaceCamp*[2]—shuttles appear less as reliable space vehicles than accidents-in-waiting. In *Starflight,* the last in a line of *Airport* disaster films focusing on the dangers of air travel, two shuttle orbiters (Columbia and a secret military shuttle) attempt to rescue the passengers of a new hypersonic jetliner stranded in Earth orbit. The jetliner, a winged space vehicle whose technology was under development at the time (and which President Ronald Reagan actually proposed to Congress in 1986), proves to be a death trap. Technologies for the on-orbit rescue by space shuttles do not work, and the passengers and crew face dire peril. In *SpaceCamp,* the shuttle is so prone to malfunction that a foolish but well-meaning robot accidentally launches a group of teenagers from the US Space & Rocket Center into orbit.[3] (On a field trip from the real-life US Space Camp, the teenagers have already received detailed instruction about shuttle systems when they are launched into space.) Ironically, the robot triggers the launch by exploiting the same vulnerabilities in the shuttle's solid rocket boosters (SRBs) that felled Challenger, and the film's release a few months after the disaster doomed it to oblivion.[4]

After January 1986, the shuttle's balky fragility and propensity for problems became its enduring identity; flying the craft during the 1990s and 2000s often meant merely keeping it flightworthy, a task shared by thousands of workers and contractors at the National Aeronautics and Space Administration (NASA), but for which astronauts bore the greatest risks. Throughout the shuttle's last two decades, a legacy of fear hovered over it, with crew members' orange Launch and Entry Suits and Advanced Crew Escape Suits (known colloquially as "pumpkin" suits) a grim reminder of the overconfidence that had plagued the shuttle's early years, when astronauts flew in blue coveralls and lacked even rudimentary protection from the space environment. Acting as both a forensic analysis of past mistakes and a warning for the future, works like Diane Vaughan's *Challenger Launch Decision* served as a constant reminder of the supposed weaknesses of NASA's management practices. Popular culture about the space shuttle after 1986 shared this newfound sense of the craft's precariousness: 1998's *Armageddon*,[5] 2000's *Space Cowboys*[6] and *Mission to Mars*,[7] 2003's *The Core*,[8] and 2013's *Gravity*[9] offered sequences of space shuttles damaged or otherwise in distress, with their crews facing imminent peril.

Despite NASA's investments in the Space Transportation System (STS), efforts to replace it with a safer or more capable architecture were continuous throughout the program. Contractors and NASA engineers offered alternatives utilizing proven hardware throughout the 1970s, while parallel shuttle developments in the Soviet Union suggested that the technologies of other heavy-lift rocket programs might improve shuttle reliability, like the use of liquid-fueled boosters instead of troublesome solids. Choosing not to proceed with this option, NASA began, instead, a twenty-year odyssey to replace the space shuttle with a spaceplane with even more unrealistic capabilities, culminating in a series of embarrassing development failures. Time after time, the shuttle escaped cancellation, not due to its fitness for its intended purpose but because replacements were unavailable.

Imagining Alternatives

Imagining the last forty years of American spaceflight without the shuttle is an exercise in alternative history that strains scholarly analysis to the breaking point. The absence of a mandate to build one infrastructure does not mean that Congress would have spent the funds earmarked for it on another, and a future without any American human spaceflight program, though improbable given the broad US consensus to sustain at least some form of human space exploration, is plausible under different political circumstances. More likely outcomes, though, may be premised on the essential role that spaceflight played in American life in 1972. That the US would retain a satellite launch capability after Apollo 17's final Moon landing, for example, was not in doubt in 1972: NASA's civilian space program had always necessitated a range of expendable launch vehicles to handle vital space traffic, including weather, communications, and scientific satellites, and the national security community enjoyed its own parallel space programs, which the shuttle affected only slightly. That the vehicles that launched these payloads would remain in service past 1972 was assured regardless of the shuttle decision, as the shuttle would take years to reach operational status and replace them.

Without the shuttle, NASA's parallel, "manned" and "unmanned" space programs (the gendered terms persisted in agency use until 1973) would likely have continued. The human spaceflight alternatives to the shuttle were numerous in 1972, and a decision not to build the shuttle would not have necessarily ended American human spaceflight. The cancellation of lunar flights after Apollo 17 left NASA with a handful of expendable Saturn IB and V launch vehicles and Apollo command, service, and lunar modules that could be repurposed for long-duration lunar or Earth-orbit missions or as scientific laboratories.[10] Initial experiments to do this under the Apollo Applications Program (AAP)—culminating in Skylab—were resounding successes, bought on the cheap, and repeatable with hardware already on hand.

Absent a decision to build the shuttle, the recession of the mid-1970s would have likely delayed the next big space program in favor of a continuation of the AAP and a slow growth in spaceflight capabilities. Crew launches, tailored for specific missions like earlier Apollo flights, would be relatively infrequent, most likely flown at the rate of two per year at 1970s funding levels, sustaining human spaceflight but not creating the renaissance—the Golden Age of Spaceflight—for which shuttle enthusiasts had hoped.[11] Whether future craft visited the Moon or not, future space hardware would remain optimized around the Apollo paradigm: expendable vehicles, small crew capsules with manual controls, and flight crews consisting primarily of aviators. The Soviet space program, even more constrained by technological deficiencies and finances, ultimately chose this path.

Without the shuttle, though, NASA's human spaceflight program would have faced a moment of reckoning by 1977 once NASA has expended the Apollo–Saturn hardware built for the lunar program. Contractor production lines for Apollo hardware would have long ceased operation by then, and while restarting them would be expensive, doing so would at least produce technology known to work well. (Apollo spacecraft might have even served into the 1980s and beyond, modernized and made safer, as was the case with the Russia's Soyuz spacecraft.)[12] To be sure, without a reusable launch vehicle, Apollo launches would have remained expensive through the 1970s and 1980s, but costs would have dropped over time. Experiments in the 1960s demonstrated that Apollo command modules and Saturn first stages could be refurbished and reused; NASA managers like George Low were advocates of such an approach, viewing Apollo–Saturn hardware as a better near-term bet for the continuation of a viable and affordable human spaceflight program. The STS, by contrast, was a luxury, easily pushed well into the future, should the US still want it then.[13]

With a surfeit of reliable new Apollo hardware in the late 1970s, the US lead in space exploration might have grown even larger. Initially ahead of the US with high-thrust rocket stage and early piloted flights, the Soviet space program stagnated with the sudden death of its principal designer, Sergei Korolev, in 1966, and spacecraft intended to complete a Soviet Moon voyage suffered repeated failures during the late 1960s as American vehicles showed increasing promise. It was only in the mid-1970s that NASA learned that Soviet space hardware lagged badly in many key design areas, sitting roughly between 1961's Project Mercury and 1965's Project Gemini in the sophistications of its vehicles. This technological superiority put NASA in a position to lead in human spaceflight even absent the shuttle.

Throughout the Cold War, American and Russian space activities were wholly independent of each other, and while both nations engaged in diplomatic relations regarding space matters, only a single joint flight occurred.[14] The product of over a decade of negotiation and hardware integration, the 1975 Apollo–Soyuz Test Project (ASTP) saw spacecraft from both nations rendezvous and dock in space for a brief crew exchange, a mission facilitated by American technological capabilities. The two spacecraft architectures had never been designed for mechanical compatibility and employed two different pressurization and gas systems in crew compartments: while the Soviet Soyuz had employed a nitrogen-oxygen mixture at sea-level pressure (14.7 pounds-per-square-inch), Apollo vehicles used pure oxygen at reduced pressure (5 pounds-per-square-inch): direct connection of the space vehicles might have caused an explosion, and crew members transferring from the high- to the low-pressurization regime would have developed the bends, potentially killing them. Success in docking stemmed from Apollo's adjustment of its orbit to match Soyuz's inclination, and its orbiting of an airlock module and mechanical adapter to connect the two craft.[15] While not used again, the technology worked well. A second Skylab, constructed as a backup in 1970, remained in NASA's inventory throughout the decade, but plans to launch it, and perhaps dock it with a Soviet Salyut station to create an

"International Skylab" were shelved to free up funding for the space shuttle.[16] The decision was a painful one: with Saturn V and Saturn IB launch vehicles and Apollo spacecraft in storage, enough hardware already existed to fly several years' worth of missions to the joint station. Instead, NASA donated flightworthy hardware to museums, and the US undertook a far more expensive plan to build a similar station decades later.[17]

NASA's lack of a specific exploration goal—and its reluctance or failure to find one during the 1970s and 1980s—might have hurt its standing with the public, journalists, and Congress in time, but this outcome would have been no worse than its experience with the space shuttle. With the exception of wary observations of space activities by the People's Republic of China (and brief concern in Congress during the mid-1980s that the Soviets might reach Mars), American human spaceflight during the shuttle era also lacked a clear mission or purpose, other than creating new capabilities.[18] Years earlier, NASA had recognized this problem, with one senior manager writing to NASA administrator James Fletcher in 1973 of his desire to "get away from the project-oriented mode,"[19] but the agency eventually found that selling the public on an agency with amorphous scientific goals was harder than pitching one that undertook straightforward campaigns of exploration. NASA appeared most successful when it set discrete goals and met them, and these goals did not require an all-purpose space vehicle like the shuttle. Major NASA goals of the shuttle era—operation with international partners and construction of an Earth-orbiting space station—could have been easily accommodated with Apollo hardware, likely for a fraction of the cost.

Snowstorm: The Soviet Shuttle

For leaders of the Soviet Union's cloistered and secretive totalitarian government—who had watched the development of the space transportation system with a mixture of puzzlement and envy—the shuttle's operational difficulties did not come soon enough to prevent them from embarking upon one of the most expensive spaceflight programs in Soviet history, and arguably its greatest mistake.[20] Like their American counterparts, Soviet leaders had canceled an earlier spaceplane: the MiG-105, begun in 1965 to replicate the capabilities of the US Air Force's X-20.[21] For the Soviets, the responses to the later US shuttle program were akin to earlier American responses to Sputnik 1 and included both an acceleration of existing military space vehicles

Figure 6.1
Launched into space atop an expendable Energia liquid-fueled rocket, the Soviet Union's Buran space shuttle and its launch complex appear in this 1986 Defense Department artist's conception, as a masterwork of brutalist architecture. Defense Audiovisual Agency photo, courtesy of NARA, National Archives Identifier: 6399402.

and the development of new ones. Like their American counterparts, Soviet engineers had found an abundance of technical problems and a lack of utility in the spaceplane concept and had terminated their original work after five years; following NASA's initiation of the space shuttle program in 1972. However, Soviet work resumed, leading in 1974 to a decision under Soviet premier Leonid Brezhnev to develop a comparable vehicle, Buran (snowstorm), the name of both the program and the first completed orbiter (figure 6.1).[22]

To Soviet leaders, the American shuttle's purpose was unclear, as only the most sinister military objectives seemed a good match for the shuttle's unusual capabilities.[23] Few civilian satellites were large enough to require a 15-by-60-foot cargo bay, and no known civilian payload needed to be returned intact to Earth: satellites were more easily replaced than repaired. Such capabilities, though, were useful if US military branches or intelligence agencies sought to quickly orbit reconnaissance satellites, lift and service exotic weaponry into space, or capture Soviet satellites for inspection and analysis. Soviet leaders preoccupied with national security concerns (like Soviet defense minister Dmitry Ustinov and Soviet Academy of Sciences president Mstislav Keldysh) feared the worst: that the shuttle would be used as an orbital weapon of some kind, either an Earth bombardment craft[24] or an orbital attack plane—hijacking, destroying, or capturing the Soviet Union's nearly defenseless satellite fleet. (Indeed, Soviet designer Vladimir Chelomey had proposed such a technology, unsuccessfully, to Brezhnev's predecessor, Nikita Khrushchev, in 1960.)[25]

In the absence of such intentions, the shuttle appeared a poor investment, and when Soviet engineers analyzed the craft, some recommended against its construction, only to be overruled by political functionaries convinced that the Americans would not have invested so much money in such a dubious project without a good reason.[26] Ironically, a lack of understanding of American domestic politics doomed the Soviet Union to expend desperately needed capital on an expensive experimental space vehicle for which neither nation had a clear use.[27] In the most controversial move, though, Soviet leaders eventually elected to duplicate the shuttle orbiters rather than improve them. As details about the shuttle's design were largely unclassified,[28] Soviet intelligence had little trouble acquiring details about its configuration and systems; studies by the Zhukovsky Central Aerohydrodynamic Institute near Moscow verified American research on the optimal shape of the craft,[29] and Buran took form under its chief designer, Valentin Glushko, as a delta-winged glider virtually identical to Columbia (figure 6.2).

For the Soviets, immediately constructing a space shuttle completely identical to the American craft was impossible; its technological and industrial capabilities were far less sophisticated. The typical decade-long gap between the appearance of an American flight vehicle and its Soviet counterpart, though, often enabled Soviet engineers to reconfigure Western designs to suit local manufacturing capabilities. The abortive American Manned Orbiting Laboratory, an Air Force space station planned by the Pentagon in the 1960s, never moved beyond initial hardware testing, but by 1974, Russia had successfully orbited a military space station of its own, Almaz,[30] and added a component not found on the American version: a 23-millimeter automatic canon, borrowed from Soviet bombers, to defend it from American attack.[31] With its technology behind that of the US, the Soviet Union built Buran beginning in 1980, with Columbia already nearly complete, but with a number of design

Figure 6.2
This 1985 artist's conception from the Department of Defense depicts the Soviet space shuttle Buran (right) docking with a Soviet military space station (left). Buran's resemblance to Columbia, achieved through direct duplication, is obvious. Defense Audiovisual Agency photo, courtesy of NARA, National Archives Identifier: 6386730.

differences to suit domestic manufacturing capabilities and design traditions. These differences provided an echo of US shuttle mode debates of the early 1970s, with the Soviets pursuing the propulsion and design options that the US had rejected.

The first sacrifice for the Buran project was full reusability: while the orbiter could be flown again, a new expendable launch vehicle would lift it into orbit: Energia, composed of a liquid-fuel core and two strap-on liquid-fueled boosters. Instead of placing main engines on the orbiter, Buran relied on engines mounted on the booster, sacrificing them with each mission but simplifying the development of the craft and its maintenance between flights.[32] Designed to minimize development rather than operating costs, Buran's efficiencies would never approach those of the Thrust-Assisted Orbiter Shuttle (TAOS), but for Soviet planners, operating efficiencies were not a concern: without civilian uses,[33] Buran would fly only on military missions, for which funds were assured. The resultant craft offers a tantalizing clue as to what the American shuttle might have looked like had NASA's aims been more limited. Buran dispensed with segmented SRBs and Columbia's reusable main engines. And with the Soviet space program less dominated by the test piloting fraternity, the American shuttle's dependence on hands-on piloting during landing was replaced, in the Soviet version, with a fully automated piloting system that enabled Buran to launch and land devoid of a crew.

Soviet faith in its new spacecraft, though, was not accompanied by the financial resources to operate it. After flying into space once, unpiloted, in 1988, Buran was permanently grounded amid the political and economic turmoil surrounding the collapse of the Soviet Union. Instead of flying Buran again and completing follow-on vehicles, the postcollapse Russian state retained the Soviet Union's 1960s-era Soyuz capsules and expendable launch vehicles. As a vehicle to usher in the next age of Soviet space achievement, Buran was a costly mistake that further undermined Soviet economic stability. While reaching a similar conclusion about its own shuttle's utility, the US continued to employ the STS while searching desperately for its replacement.

Replacing the Shuttle

In focusing its spacefaring efforts on the operation of a reusable, winged space shuttle, NASA managed to achieve a goal that had eluded it in the dozen years that followed the triumph of Apollo 11's dramatic landing on the Moon—establishing consensus about the future of human spaceflight in the US. In a staggering achievement for the space agency, virtually every stakeholder in American space exploration by 1985 shared the view that the shuttle required replacement. As Alex Roland wrote in *Discover* magazine in November of that year, the shuttle was a lemon (or, more accurately, a turkey): overbudget, overweight, and oversold, less an eagle soaring to new heights than an albatross for an agency still searching for the next big thing in space exploration.[34] With the shuttle's weaknesses already apparent, the question of America's proper post-Apollo space infrastructure reemerged and remained unanswered for the next thirty years. In 1990, the Advisory Committee on the Future of the US Space Program, convened by Vice President Dan Quayle and chaired by Martin Marietta chairman Norman Augustine, recommended deemphasizing the shuttle; by 2011, the shuttle's in-flight mishaps and NASA's numerous false starts at replacing it had forced the agency to abandon spaceplanes entirely and embrace, once again, expendable boosters, capsules, a separation of military and civilian scientific spaceflight operations, and the abandonment of private space tourism as an objective.

NASA had originally been hostile to suggestions that it undertake a phased development program for the shuttle, but eventually it chose an architecture whose chief benefit was that it would facilitate it incremental approach without drawing too much attention to this fact. As its designers at Mathematica and the Institute for Defense Analyses had originally intended, TAOS served as a test bed rather than a fully operational transportation infrastructure, and efforts to quietly upgrade the STS began almost immediately with the program's first flight in 1981. TAOS's great appeal, in fact, was the modularity of the concept and the ease with which obsolete, underperforming, or risky subsystems could be replaced with more reliable and powerful alternatives, or the orbiters augmented with features stripped from the design during the shuttle's arduous development process. Retrofitting of the shuttles with improvements continued throughout the shuttle's operation, but never achieved the complete redesign that the shuttle required, focusing instead on modest improvements in computing, cockpit displays (the "glass cockpit"),[35] and thermal protection. Having committed to the construction of an orbiter fleet flying regularly, NASA lacked the time and resources for a more comprehensive redesign; instead, the agency sought to ameliorate the most immediate of the shuttle's deficiencies. When this proved impossible, it quietly sought to replace the problematic piloted orbiters with either an unpiloted shuttle derivative or an entirely new piloted spacecraft.

During the 1970s, 1980s, and 1990s, NASA examined a variety of options to repurpose STS hardware to avoid the cost and safety problems associated with using the shuttle as an all-purpose launch vehicle. These designs varied in their degree of audacity and reuse of shuttle technology, from pilotless cargo carriers to plans for exotic new piloted spacecraft closer to Max Faget's original vision of an orbital spaceplane. The most viable concept would have stripped the shuttle orbiters of wings and crew compartments to create cargo vessels capable of delivering heavy payloads to orbit without human participation.[36] Although frequently resurrected as the Shuttle-C or Shuttle-Derived Heavy Lift Vehicle, these spacecraft designs never received significant support from NASA, despite their simplicity and safety.[37] Through the 1980s and even later, NASA's increasingly desperate marketing for the space shuttle insisted that all spacecraft functions—such as crewed flight, scientific study, and delivery of cargo for civilian and military applications—could be performed most economically by a single piloted craft. In practice, this insistence produced scheduling nightmares and disappointed scientists, and it also fueled popular opposition when NASA asked astronauts to risk their lives carrying into space satellites previously hauled on expendable launch vehicles. The projected efficiency of cargo variants of the shuttle served merely to cast doubt on the wisdom of the entire shuttle program, earning these designs no goodwill among those at NASA still determined to fly crewed spaceplanes.

Rather than trim the shuttle to its essential functional elements, the US Department of Defense, NASA, and their contractors, emboldened by the Reagan administration's enthusiasm for private enterprise and space weaponry, decided to double down on a troublesome idea and attempt an even more audacious replacement, blending runway landing and reusability with the most demanding technical challenge contemplated by the first generation of rocket pioneers: single-stage-to-orbit (SSTO) flight. To enthusiasts for this new iteration of the spaceplane, NASA's space shuttle had validated the concept of winged flight and landing; NASA had merely built the STS too early and too cheaply to capitalize upon technologies that would have made it economically viable. A single-stage, reusable winged launcher would be a true space airplane, one that would shed no components on liftoff, never require the fabrication of new tanks or stages, and might even take off from and land at airport runways.

Achieving the speeds necessary to launch a vehicle into space requires a craft so lean that propellant constitutes over 90 percent of its launch weight (see chapter 1).[38] Konstantin Tsiolkovsky's solution to this requirement was the multistage rocket: shedding empty tankage along the rocket's path by stacking rockets on top of each other. While the staging paradigm later became dominant in launch vehicle design during the late twentieth century, it offered a variety of challenges. Early liquid-fueled rocket engines reliably ignited only half the time, and before 1950, they had never been started in midair or in space; as a multistage rocket might have several engines in each stage, this degree of unreliability made a functioning staged rocket a statistical anomaly. Early engineering studies examined the SSTO as an alternative: a 1946 satellite proposal by North American Aviation (NAA) for the Navy Bureau of Aeronautics—the High-Altitude Test Vehicle (HATV)—had a dry mass of only 11 percent of its fueled weight and would have orbited itself and a small payload. Far more advanced than any subsequent space vehicle, the HATV attracted no development funding at the time.[39]

Although the SSTO found few enthusiasts in the 1940s, the unreliability of liquid-fueled rocket engines and the uncertainties surrounding staging led both the US and the Soviet Union to pursue alternatives to staged rockets, including "stage-and-a-half" designs in which all the launch vehicle's engines ignited at liftoff, with engines or tankage discarded later in flight. Sergei Korolev's launcher for Sputnik 1, based upon the R.7 intercontinental ballistic missile (ICBM), consisted of four liquid-fueled rockets strapped to a liquid-fuel rocket core with longer propellant tanks; in flight, the four boosters separated upon depletion of their fuel, and the central (sustainer) stage continued to propel the craft upward. For a launch vehicle of its size and mass, an R.7 derivative could orbit only a relatively small payload in addition to the empty tankage that it dragged into space.[40]

Another iteration of this technology was the SM-65 Atlas, an early Air Force ICBM pressed into service as a space launch vehicle. An innovative design produced by Convair, Atlas was the brainchild of a Belgian émigré named Karel Bossart, who pioneered an efficient "balloon-tank" design that used propellant tanks pressured with inert gas as structural elements supporting the missile's entire weight.[41] In lieu of the more conventional staging, Atlas employed a stage-and-a-half design, in which all three of the missile's engines ignited on liftoff, fed by a single set of propellant tanks feeding them, with two engines jettisoned shortly after launch and the remaining ("sustainer") engine firing to complete the missile's powered ascent. In 1958, only months after the launch of Sputnik 1, the Department of Defense's Advanced Research Project Agency (ARPA) demonstrated the lifting capabilities of the Atlas B by orbiting the missile, a tape recorder, and a small radio transmitter as Project SCORE (whose name stood for "Signal Communication by Orbiting RElay"). Although launched, ostensibly, to demonstrate the possibilities of satellite communication (the onboard transmitter beamed a short greeting from President Dwight Eisenhower to receivers on Earth), the true purpose of Project SCORE was to demonstrate American superiority in orbital lift capability after the Soviet launch of the 1,100-pound Sputnik 2. SCORE's Atlas managed to launch the heaviest payload into Earth orbit at the time, but most of the mass carried into orbit was the missile itself—a sustainer engine and tanks emptied of kerosene and liquid oxygen, but still pressurized with residual gas.[42]

An early, successful effort to reduce mass fraction, Atlas did not become the model for later rocket design—balloon-tank construction was challenging (unfueled rockets occasionally depressurized on the ground, destroying themselves), and the impetus for the stage-and-a-half design was the unreliability of liquid fuel rocket engines and lingering concerns about the practicality of staging. Atlas's design provided an insurance policy against engine failure: if all three engines did not ignite on the pad, operators could abort the launch. Once engine reliability improved, though, upper-stage engines that ignited in flight seemed like less of a gamble, and staging became the norm.[43]

Ironically, improvements to rocket engines in the late 1950s also made SSTO vehicles more plausible: modern estimates suggest that the most high-performance rocket stages of the 1960s—like the first stage of Titan II—could have orbited themselves if configured to do so. Apparently, no entity ever tried to undertake such a stunt, as such a craft offered no benefits in the form of payload, range, or economic efficiency over multistage launch vehicles. Early space launch activities were motivated by the need to demonstrate heavy lift capabilities for nuclear weaponry or to orbit reconnaissance or communications satellites that might weigh several thousand pounds. Even the small scientific satellites proposed by the US and the Soviet Union for the 1957–1958 International Geophysical Year would strain the capability of an SSTO launch vehicle, so the concept never attracted significant interest among spacefaring nations.

SSTOs made little headway during Project Apollo because SSTO flight to the Moon wasn't possible—no launch vehicle of that type could carry enough fuel to carry a piloted vehicle into orbit, and more fuel would be required for the translunar injection, lunar orbital insertion burn, and trip back. The Saturn V launch vehicle's power and versatility, though, encouraged flights of fancy: Boeing design studies pitched Chrysler's Saturn S-IC as a stage-and-a-half launch vehicle with a single central sustainer engine like Atlas, while the engineer Phil Bono pitched an SSTO based upon the Douglas Aircraft Company's S-IVB upper stage. Although calculations demonstrated that the S-IVB could lift a piloted Gemini capsule into orbit with ease, Bono feared that the spacecraft's lack of wings (it would land on retractable legs) made it unappealing to NASA officials determined to acquire a reusable spaceplane. His later, larger, Reusable Orbital Module-Booster and Utility Shuttle was met with predictable silence from NASA.[44] Chrysler's own SSTO capsule concept, the Single-stage Earth-orbital Reusable Vehicle (SERV) competed unsuccessfully in NASA's space shuttle design competitions.

Although the idea no doubt inspired many science fiction writers, an SSTO spaceplane remained implausible throughout the 1960s and 1970s.

Maximizing the mass fraction requires economical design, including maximizing the useful internal volume with a minimum of added mass. Spheres offered the greatest volumetric efficiencies and were easier to pressurize than were other shapes; some of the earliest Soviet space vehicles (including Vostok and Voskhod) featured spherical crew compartments, but spheres are uncontrollable in flight through an atmosphere, rendering a spherical spaceplane implausible. Conical and cigar-shaped craft offered better aerodynamic performance, but the space rockets of 1930s fiction magazines (with their large fins) and the flying saucers of 1950s science fiction (with their round but squashed crew compartments) were not models worth emulating in real spacecraft. Elaborate structural components like wings, tail fins, and disks added little to the useful internal volume of the craft and complicated high-speed flight. Many features common to high-performance airplanes, meanwhile, were too heavy to employ in spacecraft: multilayered, impact-resistant glass windows pleased flight crews but extracted weight penalties that undermined a space vehicle's performance.

The only way to orbit a winged spacecraft, virtually all designers eventually realized, was atop another launch vehicle; the final stage of a multistage rocket could be winged if engineers desired, salvaging the spaceplane concept. Eugen Sänger had embraced this paradigm first, describing launching Silbervogel on a horizontal rocket sled that would accelerate the craft to high speeds even before it left the ground. Subsequent spaceplane proposals throughout the 1940s, 1950s, 1960s, and 1970s employed a booster rocket or booster aircraft; SSTO spaceplanes remained an impractical notion relegated to low-budget science fiction film and television. Yet among the earliest proposals to replace the space shuttle in the 1980s was an effort directed at creating an SSTO spaceplane capable of carrying both passengers and cargo into Earth orbit and back. The imagined advantages of an SSTO never included low-cost heavy-lift capabilities: rather, SSTOs principally offered faster turnaround between missions and lower operating costs for passenger flights,[45] benefits of marginal use in scientific spaceflight but useful for servicing a space station, or for national security applications like space reconnaissance and orbital combat. Turning the existing STS into an SSTO would require substantial reengineering, but engineers at NASA's Langley Research Center (LRC) proposed doing just that by mounting six shuttle engines on the External Tank (ET) and fitting it with wings and a crew compartment, constructed using composite materials lighter than aluminum. The proposal went nowhere.[46]

In 1982, though, at the urging of the Air Force lieutenant general Lawrence Skantze, the Defense Advanced Research Projects Agency (DARPA, formerly ARPA) authorized funding for a project called "Copper Canyon," to develop a new winged hypersonic spaceplane. Technical advances did not drive renewed interest in the SSTO spaceplanes; rather, doubts about the space shuttle's safety[47] and President Reagan's interest in space-based weaponry presented the Air Force with an opportunity to undertake a massive procurement effort, even for space vehicles that had a minimal chance of working. Reagan's Strategic Defense Initiative (SDI),[48] announced the following year, anticipated the orbiting of large numbers of military satellites capable of shooting down incoming ballistic missiles from the Soviet Union, possibly on short notice; fielding these required routine access to orbit. When the space shuttle proved unable to provide that, the winged SSTO offered an appealing alternative.[49] Ironically, Reagan's scandal-plagued White House also employed several individuals who were central to President Richard Nixon's chaotic decision to develop the space shuttle ten years earlier, including George Schultz and Caspar Weinberger.

Copper Canyon would be powered by scramjet engines that burned liquid hydrogen and ambient oxygen, potentially halving the propellant requirements for the launch vehicle's ascent through Earth's atmosphere.[50] Vehicles powered by such engines would have the speed of rocket engines (perhaps reaching nearly twenty times the speed of sound, or about 15,000 miles per hour) but a fuel efficiency closer to commercial jet planes, provided that they stayed within Earth's upper atmosphere. The addition of a bipropellant rocket engine, or modification of the scramjet to utilize liquid oxygen carried aboard the craft, would enable such a vehicle to accelerate by the additional 2,500 miles per hour required to achieve orbital velocity, and then decelerate the craft for its return from orbit. This return would be a powered, controlled flight at high speed through the atmosphere, giving the craft its choice of landing fields, perhaps even at major urban airports.

This concept was not new, and the intervening years had not made it significantly more promising. Mstislav Keldysh and his various American counterparts had worked on such a craft unsuccessfully in the late 1940s, and Walter Dornberger had unsuccessfully pitched a similar idea at Bell Aircraft in the 1950s. The excessive performance demands of hypersonic spaceplanes required a host of nonexistent technologies, including lightweight tankage and wing structures, and rocket engines capable of efficient operation both within the atmosphere and outside it. NASA's own experiments with scramjets in the 1960s produced nothing that worked, and while LRC theorized a new technique for integrating a scramjet into an aircraft's fuselage in the 1970s, it was unable to get the new engine to function in tests by the time of DARPA's Copper Canyon decision.[51]

The loss of the shuttle orbiter Challenger in 1986 accelerated the scramjet plane's development: Reagan, in his Challenger-delayed 1986 State of the Union address that year, suggested the construction of an experimental National Aero-Spaceplane (NASP) and also raised the possibility of a new high-speed aircraft, dubbed the "Orient Express," offering high-speed passenger travel for everyday people (figure 6.3).[52] (Subsequent accounts of the NASP program often conflated the NASP with the Orient Express, the latter of which would not be capable of orbital flight.)[53] Although the earliest artists' conceptions of the NASP depicted a delta-winged airplane similar to

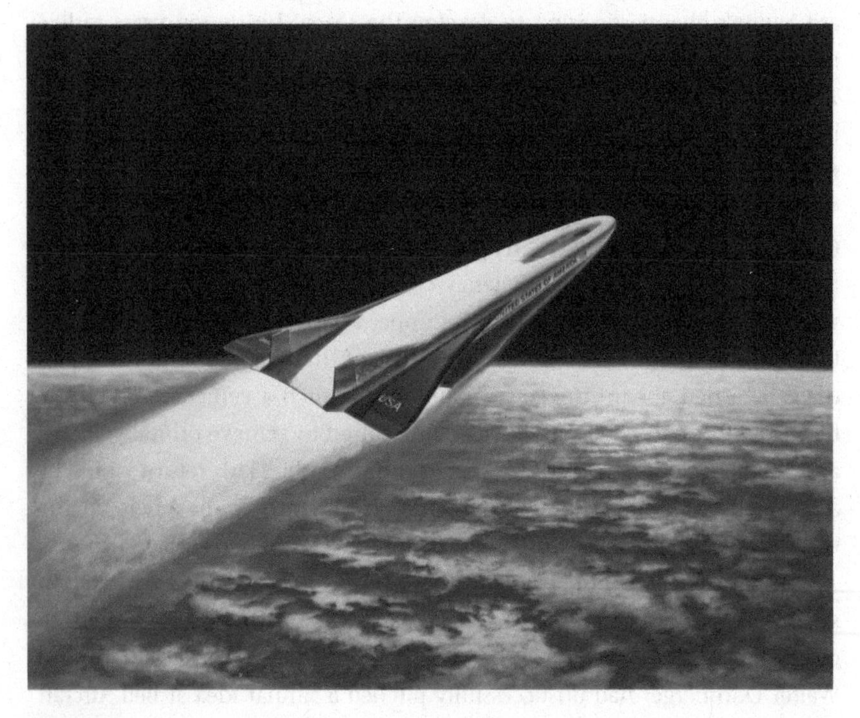

Figure 6.3
An artist's conception depicts the X-30 National Aero-Space Plane (NASP) in flight. NASA photo.

supersonic transports like the turbojet-powered BAE Concorde, the weight, thermal, and structural problems associated with wings pushed NASA toward lifting bodies—another embryonic technology of the 1960s that the agency had earlier rejected as impractical. Eliminating wings from gliding designs was appealing for several reasons: in every phase of spaceflight except landing, wings were a hindrance whose weight, high drag, and structural vulnerabilities complicated spaceplane design. Even eliminating wings was insufficient to achieve the necessary mass fraction, though, and after lightening the craft, engineers were so short of options that they decided to freeze the liquid fuel into a dense slush in order to fit more molecules of it into the tanks.

Over the next six years, a variety of established aerospace contractors received funds—approximately $250 million a year—to develop the NASP (dubbed the X-30), a significant sum but only a fraction of the estimated $15 billion and ten years required to produce a functioning vehicle.[54] While the decline in military spending associated with the end of both Reagan's presidency and the Cold War affected the pace of development, it was ultimately the presence of too many overwhelming technical challenges—in engines, heat management, tankage, and structures—and other issues that led to the X-30's cancellation in 1993.[55] Despite the NASP's demise, though, NASA's enthusiasm for an SSTO spaceplane continued. Instead of imagining a craft making a gliding, hypersonic ascent through the atmosphere, the agency instead turned to a hybrid SSTO spacecraft that would launch like a rocket but might employ lift-generating structures to slow its return and descent.

In 1994, NASA initiated a new program, dubbed the "Reusable Launch Vehicle," which solicited additional proposals from contractors for shuttle-like space vehicles.[56] The resemblance of this solicitation to those that had produced the original space shuttle in 1968 was eerie: again, the same mix of reusable winged craft, lifting bodies, and the odd capsule vied for acceptance. A truly reusable, low-maintenance SSTO vehicle would offer cost economies whether winged or not, and several American firms offered vertical takeoff and landing (VTOL) capsules and pods that were also reusable.[57] Support for such a craft had originally come from the SDI,[58] but having already settled on a winged vehicle, NASA was unsympathetic toward VTOL proposals, including the DC-X, McDonnell Douglas's legged, vertically launched capsule. Again, institutional biases prevented NASA from accepting a nonlifting vehicle configuration. Ignoring the DC-X was especially easy for NASA, as other contractors rushed to provide the agency with exactly what it was looking for instead.

Lockheed's proposed VentureStar promised a large liquid-fueled launch vehicle in the form of a lifting body with a mass ratio sufficient to enable vertical liftoff, orbital flight, and landing on conventional runways. (A certain resemblance to Lockheed's early Star Clipper was evident.) In many ways, VentureStar would have been both a correction of the original shuttle design and a more practical version of the NASP, offering a single, integrated fuselage and propulsion system, metallic heat shield, and greater cost-effectiveness. Like the shuttle, though, it conflated lifting flight with a variety of other desirable flight qualities, including reusability and economy of operation.

Instead of pursuing a VTOL craft, NASA proceeded in 1996 with the development of a subscale version of the VentureStar, the X-33, which would be capable of only suborbital flight but which would validate the design of its larger, orbital sibling.[59] Again, NASA chose a flight architecture based upon a single preferred design characteristic—runway landing—to the exclusion of other alternatives offering similar cost structures. Having compromised on design with the shuttle, NASA was unwilling to do so again, but the addition of SSTO capability to the already challenging demands of hypersonic flight proved the project's undoing. After seven years of design work, $1.3 billion in funding provided by NASA and Lockheed (in a 3-to-1 split), both entities realized that they had underestimated the technical challenges, particularly in the construction of lightweight fuel tanks for the mass-sensitive vehicle. When the fuel tank of the subscale X-33 technology demonstrator exploded in a ground test, the program lost the confidence of its supporters, leading to its cancellation in 2001.[60] Twenty years into the shuttle's operation, NASA was no closer to a replacement vehicle than it had been when the program started.

The following year, another space shuttle follow-on met a similar fate. In 1995, NASA undertook the development of a new lifting vehicle intended to supplement or replace the shuttle in ferrying crews to the International Space Station (ISS). NASA intended the X-38 Crew Return Vehicle to serve as a kind of "lifeboat" permanently docked to the ISS, capable of returning the station's entire seven-person crew to Earth in the event of an emergency. The craft, an enlarged version of the X-24A lifting body that NASA had tested in the 1960s, would not contain a cargo bay; rather, it would glide to a runway landing like the shuttle, offering a larger crew capacity and a safer mode of reentry and recovery than the three-seat Russian Soyuz capsules then in use as station lifeboats. Mounted atop a suitable expendable launch vehicle, the X-38 might even serve as a replacement for the space shuttle as an ISS ferry vehicle. Although promising, the X-38 fell victim to budget cuts one year after the X-33, leaving NASA again with no means of visiting the ISS beside the shuttle, and no craft on the drawing board to replace it. (In 1992, the European Space Agency ended development of a similar French space glider, Hermes, which would have launched atop an Ariane 5.)

Prior to the destruction of Columbia, commentators had predicted that the loss of a second shuttle orbiter would likely ground the STS program, speculating that an existing expendable launch vehicle developed through the Air Force's Evolved Expendable Launch Vehicle program (most likely a Boeing Delta IV or Lockheed Martin Atlas V) (in 2006 the companies formed the United Launch Alliance to manufacture both launch vehicles), would carry one of the new lifting space vehicles under development (most likely an X-38). The inability of NASA and its contractors to produce a viable replacement craft for the shuttle by 2003 left these hopes unfulfilled, leaving the agency with only two unenviable choices: abandoning the shuttle program (and completion of the ISS) or continuing to operate the shuttle in a restricted manner to complete the most pressing STS program objectives within newly accepted standards of risk. It chose to do the latter, and while these efforts proved successful, they inevitably exposed NASA to criticism that it had taken, simultaneously, both unnecessary risks and an insufficient number of them to accomplish its goals. Most pressing in these decisions was NASA's need to complete, for its own benefit and that of its international partners, the ISS, the first component of which had launched before the loss of Columbia, and which, like the Hubble Space Telescope (HST), was tied to continued space shuttle operations.

NASA's final effort to replace the shuttle similarly lasted only a few years before being unceremoniously dumped. Constellation, a program intended to replace the shuttle with an STS-derived expendable heavy lift vehicle and a capsule 40 percent larger than that used in Project Apollo, suffered similar dissonance between plans and budgetary allotments, resulting in its cancellation in 2011. A modification to this plan employing the same Orion capsule and a greater reuse of shuttle launch technology, dubbed the Space Launch System (SLS), became the next viable replacement for the shuttle that year. The technology represented a complete abandonment of the shuttle's flight mode and unique capabilities, including the shuttle's embedded cargo bay and satellite return capability. After forty-five years of development with the space shuttle, NASA had returned to where it had begun. Among the few features retained from the STS for SLS was segmented SRBs, among the most controversial elements of the TAOS shuttle design.[61]

Conclusion

Few space vehicles have inspired as much speculation as to their alternatives than NASA's space shuttle. A compromise among competing designs, it proved inferior to both more- and less-expensive vehicles and inspired decades of second-guessing. The rich alternative history on postlunar Apollo spaceflight[62] hints at the actual work undertaken by many NASA and contractor engineers to prove that existing architectures could have met American space goals of the 1980s and beyond. The troubled history of the shuttle bears at least some of this out: while Apollo hardware could never usher in an age of cheap, reliable space exploration, neither could the space shuttle, and legacy hardware might have, as Robert Truax (1971) had noted, accomplished "90% of the real goals at 20% of the cost without really stretching the state of the art one bit."[63]

The Soviet experience with Buran suggests that the shuttle was an expensive architecture to fly no matter which nation operated it, and the inclusion of a human crew in the NASA version had been unnecessary. Finding little use for its own shuttle, the Soviet Union retired it as soon as was practical. The American space shuttle never caused the kind of economic dislocation that Buran inflicted on the Soviet state—not because the American shuttle was much cheaper to operate, but because the American economy could absorb the shuttle's growing costs.

That NASA's shuttle had, in a larger sense, failed was an open secret within the agency by the mid-1980s, and work was already underway on replacement vehicles that might move the agency closer toward its dream of cheap, routine spaceflight. The next decades, though, dealt NASA two seemingly crippling blows: another catastrophic mishap stemming from faulty design and the failure of the shuttle's intended replacements. These replacements had been cursed with some of the handicaps that had doomed the shuttle: excessive ambition in the face of physical reality, a determination to reduce costs beyond what was prudent, and problematic relationships with the Air Force, whose goals for the vehicles differed markedly from those of NASA and which did not fear public scrutiny of its missteps. The predictable result of these complications was failure, not of imagination, but of reason.

Epilogue

Inspired by Nazi rocket planes, created to spy on the Soviet Union, and pitched as a tool of environmental protection, the space shuttle was an artifact of a confused and chaotic political age. Like other big technologies of the period, the shuttle epitomized the crisis- and consensus-driven nature of twentieth-century American politics and the tendency to build large systems by committee. It also formed a living, working embodiment of what historians of technology describe as path dependence: a technical solution to the problems of a particular era that persisted long after those problems had faded. Designed during the Cold War, the space shuttle flew most of its missions long after the Soviet Union collapsed; built as a military tool, it spent most of its career engaged in scientific research. Promised to be the craft that would democratize space exploration, it developed a reputation as being too dangerous for average people to fly. As a human spacecraft, as a partially reusable experimental vehicle, and as a heavy-lift launcher, the shuttle was a palpable (if pricey) success; as a transformative vehicle to change the paradigm of space exploration, though, it was a stunning disappointment. A craft restricted to low Earth orbit, it remained there for three decades, circling to meet a Mars ship that never arrived.

In retrospect, the idea of that the shuttle would work as intended seems optimistic to the point of foolhardiness, and indeed, it did so to many, if not most, experts at the time of its creation. The craft had always been a paradox: a spacecraft that could fly like an airplane into places where airplanes could not go—outer space, or even into orbit around the Earth. At these altitudes and speeds, the air-breathing internal combustion engines that powered all existing airplanes seized up; rockets of various kinds offered the only potential propulsive mechanism for airplanes operating at such altitudes. Even rocket propulsion, though, could barely provide the necessary energy: by adding lift-generating wings, space vehicle designers might make the craft's ascent and descent more comfortable for crews, but wings were useful for only a fraction of the spaceplane's journey (and, even then, they might be more of a hindrance than an advantage). The mass and drag of the wings also complicated the problem of accelerating and decelerating space vehicles: achieving the velocity and altitude required for orbital flight and then absorbing the frictional heat of reentry into the atmosphere. Dead weight while in orbit or in space, wings were difficult to keep cool upon reentry, and became useful again only when a spacecraft plunged low enough in the atmosphere that its wings again generated enough lift to glide. Building a winged vehicle capable of navigating these high-temperature, high-stress flight regimes represented a sizable engineering problem in the 1940s, and even the earliest work by theorists suggested that the airplane was not the proper model for future spacecraft.

The most surprising element in the story of the shuttle may be the success that a relatively small community of spaceplane enthusiasts achieved in convincing major national governments to build these craft. Entranced by the idea and undaunted by the rocket-boosted spaceplane's technical demands, the US Air Force and its contractors lobbied extensively in the 1950s for this technology, attempting to justify it with an amorphous and often fraudulent repertoire of supposed benefits. Each of these iterations—BoMi, RoBo, and Dyna-Soar—sought unsuccessfully to compensate for weaknesses in previous designs, like lack of range, lack of payload capacity, and lack of mission. Despite being repeatedly rejected as untenable, the spaceplane promised to anchor space flying firmly in the theory and practice of aviation, where a well-defined community of engineers and pilots in the Air Force, the National Aeronautics and Space Administration (NASA), and private industry already existed. Shelved by NASA in the 1960s in favor of space capsules, the spaceplane concept found new life after 1968 as a civilian program to follow Project Apollo, offering the promise of an indefinite extension of the human spaceflight program facing budgetary collapse.

Neither launching a large Earth-orbiting space station nor building a Mars ship required a winged orbital space plane, but shuttle advocates within NASA pressed their case regardless, and busied themselves not only with justifying the new craft, but also with debating the countless variants that designers proposed. Most of these proposals conflated the one performance goal that NASA *needed*—economy of operation—with features that it only *wanted*: lifting flight and reusability. NASA's early invitations for proposals for what became the space shuttle excluded a specific call for a lifting vehicle, but they did emphasize reusability as the principal system goal. In practice, most contractors recognized that NASA wanted a craft with traditional wings and was unlikely to support anything else.

Each approach offered slightly different solutions to the technical challenges and cost structures—either massive upfront research expenditures or burdensome operational costs spread out over the life of the program. In gathering support for the space shuttle, NASA's conception of the craft changed until repeated compromises undermined the vehicle's reason for existing in the first place. These changes created dissension from both those who thought the shuttle was funded too much and those who thought it was getting too little. The Thrust-Assisted Orbiter Shuttle (TAOS) configuration won favor because it pushed this difficult decision into the future while appearing not to do so. While it may be tempting to call the space shuttle a success at giving the appearance of a viable space program, its selection focused innovation on technological dead ends, curtailed legacy human spaceflight activities that might have been more productive, drew money from more worthy unpiloted programs, and confined American space activities to missions in low Earth orbit instead of the planetary adventures that NASA promised.

In the absence of a real notion of how or why TAOS would be used, fantasy replaced fact. NASA sought the Air Force as a partner, hoping that its endorsement would give President Richard Nixon's administration a reason to fund it, even though the Air Force was less than optimistic. Enthusiasts for the commercialization of spaceflight (or, rather, for the government subsidy of space vehicle manufacturers) saw a cash cow—one that might enrich them at taxpayers' expense. Conservatives believed, perversely, that the shuttle would transform NASA from a money-losing government agency into a profitable, publicly owned business, returning money to the US Treasury by monopolizing the international space launch business. Envisioned in a clash of wistful fantasies and pragmatic needs, constructed on the cheap, and flown perilously for thirty years by a generation of people who lived and died in increasingly anonymity, the shuttle epitomized efforts to maintain American preeminence during the late twentieth century despite nagging fears that its best years were behind it.

In the space shuttle, NASA had vowed to continue a controversial "manned" spaceflight program but had failed to satisfy either the scientific community lobbying for more scientific probes or the growing chorus of thinkers who expected the Moon to be a practice run for interplanetary voyages.[1] As the disparity between NASA's actual plans and the public's increasingly unreasonable fantasies grew, a space counterculture emerged in opposition to the government policymakers whom it regarded as having missed a chance to transcend the mundane. (This disappointment was epitomized, perhaps, by the relegation of astronauts to servicing duties for privately owned satellites, recovering derelict communications relays for refurbishment and relaunch; see figure 4.8 in chapter 4 for a facetious take on this notion). To a sizable community of space professionals and enthusiasts, the shuttle was a wrong turn that squandered America's Apollo-era technological lead and surrendered capabilities that would not be re-created for decades. Without the political will to build a shuttle that could actually fulfill NASA's goals, the US was left with a space infrastructure that gave the appearance of functionality but was only sufficient to fulfill basic objectives and frighten the Soviet Union into building a shuttle of its own. It operated in this capacity for decades, until the Columbia disaster and the shuttle's crippling economics forced President George W. Bush to terminate the program in 2004 and for President Barack Obama to end flights in 2011.[2] Simultaneously unjustifiable and irreplaceable, the shuttle outlived every other American spacecraft and several programs intended to replace it. Its future as blank as the ceramic tiles that lined its wings and fuselage, the craft became a projection screen for the national mood about spaceflight as it ranged from enthusiasm to horror to weary resignation.

End-of-Flight

NASA's challenge, in the years following the losses of both Challenger and Columbia, was to find a use for a spaceplane that was too dangerous to provide routine access to space. Among the earliest justifications for the development of the space shuttle was the construction of a space station; by the end of the shuttle program in 2011, the need to complete the International Space Station (ISS) provided NASA with its most compelling reason for the continuation of the space shuttle program. President Bush's 2004 announcement ending the program conditioned the termination of the Space Transportation System (STS) on the successful completion of the ISS,[3] and the station was inextricably tied to the shuttle's final years, defining its purpose and highlighting its limitations.

Calls to construct a new American space station were frequent in the years following Skylab's demise in 1979, but funding seldom followed. Merely assembling a space station in orbit from components without the benefit of Skylab's mammoth Saturn V launch vehicles required a significant fraction of the shuttle's total flights, as well as providing a regular transportation infrastructure for crew transfers. The purposes to be served by such a station were always somewhat hazy, ranging from its use as a base for future interplanetary flight to a scientific laboratory, or a site for the manufacture of specialized materials using the microgravity conditions available only in space. Always eager to bolster his political fortunes by touting space achievements, President Ronald Reagan, in his 1984 State of the Union address, committed his support to an orbital outpost that shuttle crews would construct and visit by in the 1990s, but the necessary funds were not forthcoming. Originally dubbed "Space Station Freedom," Reagan's station endured protracted delays and reconfigurations until emerging, eventually as the ISS.[4]

Before then, shuttle crews practiced space station construction techniques, eventually docking with the Russian Mir space station in 1995. While the shuttle's sea-level pressurization and cabin gas mixture made renewed joint operations with the Soviet Union possible, it was political rather technological change that spurred the move to resume them. The Soviet Union's dissolution into the Commonwealth of Independent States in 1991 turned the Russian space program from a competitor into an ally; the frosty relations between the two superpowers abated (if only temporarily), and concerns about the theft or sabotage of American space technology subsided. Russia's declining economic health and capacity for self-governance instead presented international security dangers, especially the prospect, in the early 1990s, that the former Soviet Union's vast army of unemployed military missile and spaceflight engineers might ply their trade in the global weapons industry. Hoping to draw Russia's design talent away from Iran and North Korea, American politicians moved to embrace the Russian space program as a. Within a few short years in the early 1990s, human spaceflight became a new kind of diplomatic tool—one aimed not at enticing the nonaligned world away from communist influence but rather propping up former adversaries with promises of money and a shared human presence in space.

American visits to Mir were not trivial in their technical requirements. Mir (whose name translates as "world," "peace," or "village")[5] was a Soviet-era space station assembled in orbit (sometimes called a "100-ton Tinkertoy") beginning in 1986, designed to operate with the Soviet Union's Soyuz capsules and Buran. Interoperability of US and Soviet spacecraft had been a low priority after the Apollo–Soyuz Test Project (ASTP), and visits by the STS to Mir required the addition of a new docking module, which the space shuttle had ample ability to carry.[6] Atlantis's 1995 visit to Mir was the first of nine American flights to the station, and it demonstrated the shuttle's potential not only as a means by which to convey astronauts to and from orbiting space laboratories (one of the shuttle's original purposes) but the first joint US-Russian space mission since the ASTP flight in 1975.[7] Shuttle visits to Mir—including crew exchanges and long-term stays—provided useful lessons in the design and operation of large space stations, but astronauts who visited Mir recalled a workplace fraught with frequent maintenance and safety issues, including incidents in 1997 involving uncontrolled combustion and a hull breach.[8]

Mir was, to American examiners, geriatric space technology kept aloft by Russian unwillingness to change technologies that appeared to work: the Russian spacecraft that serviced the station were Soyuz capsules, first flown in 1967. Technology aboard the station was similarly behind that of the US, and Soviet engineers had committed a variety of technical faux pas, including running power cables through pressure hatches that cosmonauts would need to seal in the event of emergencies. Indeed, the principal accomplishment of the joint program was political rather than technical: demonstrating the feasibility of international cooperation and ensuring the survival of the Russian space program through an influx of American funding.[9] On balance, though, the experiment provided NASA with its only experience in conducting long-duration space flights with an international partner and provided an environment for American astronauts to stay in space significantly longer than the two-week missions possible on shuttle flights. Such missions yielded valuable data regarding the adaptation of human physiology to the space environment—a critical concern for interplanetary space flight, even if the station itself offered little promise as an example of future habitable environments in space.[10]

The most significant technical achievement of the shuttles' visits to Mir was a proof-of-concept for the creation of the ISS. A memorandum of understanding between NASA and its Russian counterpart, Roscosmos, established the joint venture in 1998, imagining a joint staging base for interplanetary flights.[11] The station, though, was unlike the Americans' previous effort,

Skylab (which was launched intact), and more like Mir: a collection of pressurized modules, trusses, airlocks, docking ports, and solar panels stretching out as large as a football field, but painstakingly assembled over a period of years by successive launches. Crew members—as many as seven in early projections—would, as they had for Skylab, Salyut, and Mir, arrive periodically in American shuttles and Russian Soyuz spacecraft, staggering their yearlong visits to maintain a permanent human presence on the station. As planned, Russian, American, and eventually European launches slowly added habitable space, solar panels, and consumables to the station throughout the early 2000s, and the construction of the ISS dominated the shuttle's final decade. The STS was one of the only launch vehicles large enough to carry the heaviest components of the station, but the necessary launches took years longer than anticipated. While the ISS managed to achieve long-duration (if not entirely comfortable) occupancy, though, significant scientific work proved elusive, and the station's human presence ultimately became the ISS's principal reason for existing, especially when delays and problems reduced the ISS's permanent crew to three.

The loss of Columbia in February 2003 at the conclusion of the STS-107 mission jeopardized the completion of the ISS: the shuttle was the only American launch vehicle capable of delivering the station's largest components, and the station was designed around the expectation that the shuttle would support it through its construction. Not only did the pause following the accident delay needed assembly missions, but NASA, in 2004, announced the phased retirement of the shuttle in anticipation of its replacement with a safer vehicle. The final seven years of shuttle operation, thus, became an exercise in triage: allocating the number of essential mission goals over a shrinking number of anticipated flights. In the final years of shuttle operation, the ISS played an additional role for the shuttle: acting as an emergency shelter for the craft in case the shuttle orbiters were so damaged upon launch that reentry and landing were impossible. Originally imagined as a lifeboat for the ISS, the shuttle had found its role in relation to the space station reversed: the ISS was now the shelter from which the crew of a stricken shuttle would seek shelter until returned to Earth by another spacecraft (figure E.1).

The shuttle's role in assembling and servicing the ISS ensured the continued operation of the three remaining orbiters—Discovery, Atlantis, and Endeavour—long after safety concerns argued for their replacement. NASA had intended the first one, Enterprise, to serve as an orbital vehicle, but

Figure E.1
The space shuttle Endeavour (left) orbits docked to the ISS (top, center) in this August 2007 photograph by an STS-118 crew member. The SPACEHAB module is visible in the orbiter's cargo bay, as is the shuttle's Remote Manipulator System. The ISS's own Candarm2 is visible at left. Much of the end of the space shuttle's flying career was spent completing construction of the ISS. NASA photo.

discrepancies in design between Enterprise and the next orbiter, Columbia, proved so significant that NASA eventually scrapped its plans to complete Enterprise as a spacecraft. (NASA found it cheaper to build a new orbiter from scratch than to retrofit an existing airframe.) Discovery was the first orbital vehicle to be retired, followed by Endeavour. Atlantis, configured as a rescue vehicle for Endeavour's final flight, launched with a skeleton crew weeks later, completing the 135th flight of the STS in July 2011—a mission to deliver essential equipment to the ISS. Ironically, noted NASA chief historian Bill Barry, the availability of the Russian Soyuz spacecraft to resupply the ISS in the short term likely hastened the decision to cease shuttle flights.[12]

The end of the shuttle operations left NASA with an enormous amount of space hardware, much of it still flightworthy and potentially useful in future space programs. With NASA leadership moving aggressively away from the shuttle in the 2000s, though, relatively little impetus existed within the agency to save equipment not expected to be used for the subsequent Constellation program. While the engines of the flown shuttle orbiters were still serviceable and could be cannibalized for testing or future craft,[13] the airframes seemed unlikely to have a future flight or research use. As a result, upon the conclusion of shuttle operations, NASA opened a vigorous competition among space facilities, museums, educational institutions, and municipalities to acquire orbiters for public display. The rarity and the tremendous cost of the orbiters' manufacture (approximately $2 billion each) made them difficult items to donate to museum collections: as taxpayer-funded technology (some of which was still sensitive), they could not be given to private individuals or foreign entities. Neither was NASA willing to allow the orbiters, as symbols of American achievement, to rust in an open field, as had Russia's Buran. The number of municipalities and institutions able to properly maintain a shuttle orbiter indoors was fewer than the number eager to receive one, and the competition to house the orbiters for display soon narrowed.[14]

Although built for and operated by NASA, the shuttles were not its permanent property: under a 1967 agreement, the Smithsonian Institution enjoyed a right of first refusal on all the space hardware that NASA chose to retire.[15] Under this arrangement, the Smithsonian, a public-private institution established by private philanthropy in the nineteenth century but later partially supported by Congress, received the most valuable American space artifacts, including the Apollo 11 command module and the spacesuits worn by astronauts on the Moon. In keeping with its arrangement with NASA, the Smithsonian Institution's National Air and Space Museum (NASM) received the first retired shuttle orbiter, Enterprise, in 1985, following the conclusion of gliding tests and a worldwide public relations tour. Permanently retired, Enterprise joined an assortment of contractor and NASA mock-ups and test articles, including Rockwell's Inspiration and the Marshall Space Flight Center (MSFC)'s Pathfinder, in becoming the first orbiters to end up as museum pieces. Enterprise, the closest of these to an actual spacecraft, eventually settled in NASM's Steven F. Udvar-Hazy Center near Dulles Airport, fitted with mock-ups of the shuttle's engines and thermal protection system (TPS).[16]

Throughout the competition to locate homes for the remaining orbiters, NASM personnel noted that the choice was not actually NASA's to make; rather, all the orbiters should have been retired to NASM's custody for display or loan.[17] Although likely correct on the merits of its argument, the Smithsonian did not press its claim, and while it received the oldest and most historically significant of the orbiters, Discovery, the others went elsewhere, including Enterprise, which left NASM's collections for the Intrepid Sea-Air-Space Museum in New York City.[18] Endeavour found a home at the California Science Center, in Los Angeles, and Atlantis, the last to fly, joined the public collections of NASA's Kennedy Space Center (KSC) in Florida, a popular tourist attraction throughout the Moon Race and the decades that followed.[19] Notably absent from the list of sites were several NASA facilities that played important roles in the program, including NASA's Johnson Space Center (JSC) in Houston; NASA's Dryden (later, Neil A. Armstrong) Flight Research Center at Edwards Air Force Base, California; and a number of popular museums, including the Adler Planetarium; the National Museum of the US Air Force in Dayton, Ohio; the Seattle Museum of Flight; and the US Space & Rocket Center. Balancing assessments of the institutions' ability to maintain and display the orbiters (with an estimated $28.8 million funding commitment) with likely public access proved most vexing for NASA selection committee members, leaving more remote institutions with close associations with the program unhappy at the shuttles' final disposition.[20]

As museum displays, the space shuttle orbiters proved extremely successful, drawing crowds and provoking the visceral awe in person that they were seldom able to on television screens. NASA's requirement that the shuttles be placed indoors made them sublime museum objects as large as a passenger jet (figure E.2). No longer subject to rigorous preflight protection, shuttles could be displayed publicly, up close, with museum visitors able to walk around and even under a craft, and even touch noncritical systems like landing gear tires. The transformation of the orbiters from dangerous craft to ancient objects of veneration, though, is still in its infancy, and as scholars continue to plumb their meaning amid a period of uncertainty in the future of American human spaceflight, the orbiters exist, uneasily, as both objects of technological wonder and artifacts of an archaic, increasingly inscrutable past.

Assessing Risk

In choosing to develop TAOS in 1972, NASA succeeded in creating a craft with the external appearance of a reusable space vehicle. There is no evidence that analysts at Mathematica or the Institute for Defense Analyses who supported the concept ever suggested that such a craft would serve as more than a demonstration of the technology: its vulnerabilities were evident from its inception. Various NASA managers mulling the design wrote to each other as if they

Figure E.2
This striking image taken at NASM's Udvar-Hazy Center on April 19, 2012, depicts the exchange of the museum's previous shuttle orbiter, Enterprise (left), for Discovery following the conclusion of the shuttle program. Enterprise was transferred to the Intrepid Sea-Air-Space Museum in New York City. NASA photo.

assumed that the shuttle would operate only briefly in that form, with its SRBs and TPS replaced with safer alternatives once the system became profitable. In effect, NASA covertly embraced the phased development program that it formally rejected, pinning its hopes not on congressional allocations, but rather on the ability of TAOS to make enough money to pay for its own upgrades.

It is doubtful that the engineers who struggled in the late 1970s and early 1980s to assemble a shuttle that could meet NASA's cost objectives expected the compromise vehicle to be completely safe, economical, or even in the interests of the US (indeed, Diane Vaughan's 1996 book *Challenger Launch Decision* suggests otherwise). Attention, during the early years of shuttle development, remained focused instead on an evolutionary approach despite NASA's protestations: perfecting certain aspects of the system (like the orbiter) while using some legacy hardware to reduce costs. Some shuttle enthusiasts at NASA recoiled at phased development throughout the 1970s, fearing (correctly as it turned out) that given the agency's tight budget, the first draft of the space shuttle would likely be the last, and innovations not integrated early into the STS would never arrive. Throughout this period, though, other NASA insiders repeatedly assumed that upgrades would eventually occur, and the shuttle would remain a dangerous machine without them.

With the shuttle never quite as cost effective or safe as desired, NASA could say that it was versatile, containing a large crew compartment and cavernous cargo bay. While the shuttle could not leave low Earth orbit, its literal emptiness meant that it could be a vessel in which to contain whatever national aspirations next proved pressing: scientific, military, commercial, or popular. NASA spent the better part of the next four decades attempting to fill the hole that it had created, with reconnaissance satellites (stewarded into space by all-military crews that compromised NASA's peaceful facade); commercial satellites (endangering the nation's space heroes with delivery duties more safely undertaken by expendable rockets); scientists (in the Spacelab laboratory module nestled in the shuttle's cargo bay); and tourists, including foreign dignitaries and private citizens. With the shuttle able to hold a maximum of eight crew members, NASA made a point of filling every seat, crowding the vehicles and, to the chagrin of astronauts, overtaxing each shuttle's lone toilet.[21]

What the shuttle did do well, at least for a time, was to launch satellites into space and recover them: these activities, NASA had hoped, would allow the shuttle to turn a profit by performing space launch and retrieval activities for private corporations. As a means of launching communications satellites into space, however, the shuttle quickly proved far too expensive and prone to delay. Having witnessed unpiloted rockets carry satellites into space for decades, few in the press could see the reason for what commentators routinely described as a "space truck" that risked human lives to do the same thing—or worse, sold astronauts' lives cheaply to earn money from long-distance telephone companies.[22] The shuttle astronauts—reduced in the public imagination from heroes to couriers—became fungible, numerous, and often depressed at their professional prospects, as indistinct to the public as the five nearly identical orbiters that they flew.[23]

In time, the space shuttle may be remembered less for what it accomplished than for the challenge involved in keeping it flightworthy. As much as its creators hoped that the shuttle would be a reusable workhorse, a surprising degree of craft knowledge held it together. Space shuttle components (particularly its rocket engines) operated under unprecedented pressures and temperatures; in addition to being monitored during flight by the factory in California that made them, the engines were literally taken apart and reassembled before each flight, a concession to safety that all but eviscerated any hope that the shuttles would fly frequently. Between missions, unwritten knowledge about how the shuttle worked replaced checklists and manuals. The shuttle's refurbishment and reuse required an army of skilled technicians for whom the operation of all aerospace vehicles was a matter of *bricolage* (tinkering)[24] and know-how passed from generation to generation during the program's forty-year history.

That the spacecraft in which the shuttle astronauts flew might fail catastrophically, killing them, was a risk that they accepted throughout the years of the program, to a degree that would likely surprise many average Americans. Indeed, the early years of the Space Race were awash in death, claiming the lives of one in four astronauts during a particularly bad stretch from 1964 through 1967.[25] As pilots with military training and, in many cases, active service during wartime, astronauts drew a distinction between calculated risk and foolhardiness, and they accepted that however carefully they might prepare for their missions, the activities in which they engaged would, by necessity, claim some of them. Flying accidents, launchpad catastrophes, and inflight emergencies were a necessary part of exploration, they believed, and the least worried among the astronauts merely believed that their superior skills would insulate them from deadly errors that might claim lesser men.

As concerns about the safety of the shuttle manifested in accidents, NASA's commitment to use the shuttle to maintain the Hubble Space Telescope (HST) and assemble the ISS ignited renewed controversy over the level of risk that the US was willing to bear to sustain its human spaceflight program. NASA's space shuttle was arguably safer than Apollo in many respects, and certainly more than the experimental Mercury and Gemini vehicles, each of which suffered their share of nonfatal mishaps. While NASA never regarded any of the earlier space programs as routine, the agency was quick to describe the shuttle as an inherently safe infrastructure, comparable to a passenger plane, in which anyone could fly without undue risk. Ultimately, this bit of marketing puffery proved unexpectedly damning once public perceptions of the shuttle inevitably met its actual safety record: good for a spacecraft, but horrendous for anything that the public might use.

Given the configuration of the STS, fatal accidents were inevitable, but the shuttle's failure rate—two total losses in 135 flights—compares very favorably with other human spacecraft for which more system risk was allowed (including Project Apollo's one fatal accident and one major, nonfatal abort in 26 flights between 1967 and 1975). The shuttle was not, as spacecraft go, peculiarly unsafe; NASA and its contractors merely oversold it as a new kind of technology offering benefits that it did not (and never could) possess, regardless of how diligent its managers and engineers were in locating and ameliorating system risks. Assigning blame to those responsible for managing and maintaining the craft, though uncomfortable for the agency, deflected attention from a more troublesome reality: the likelihood of astronaut fatalities was programmed into the STS, and there was ultimately no way to make the shuttle reliable. Veteran astronauts studied and stewed over the shuttle's deficiencies, while an ambivalent public ignored them until they became front-page news. The huge crew sizes of shuttle missions compared to previous craft—as many as eight crew members, compared to three for Project Apollo, two for Project Gemini, and one for Project Mercury—helped to ensure that inevitable disasters would outstrip all previous American spaceflight fatalities in terms of number of casualties.

The arrival of the space shuttle ultimately compromised NASA's reputation for technical professionalism, not because the shuttle was more dangerous than other space vehicles, but because the agency often tried very hard to claim that they were *less* so. Only Challenger's televised destruction in 1986 and the death of America's first space tourist, Christa McAuliffe, ended this fantasy. What had made the astronauts' sacrifice all the more grating was the lack of any real reason for it. The missions that had claimed the lives of the Challenger crew had not required the presence of so many people aboard: McAuliffe had flown merely because, in NASA's estimation, she could. The shuttle's two loss-of-crew events, in 1986 and 2003, presented NASA with contingencies for which it had planned for decades; remarkably, neither led to the cancellation of the shuttle program. Instead, aided by a sympathetic public response and effective image management by NASA[26] and the executive branch, the lost astronauts of Challenger and Columbia became martyrs to American technological greatness, with the nation honoring their loss without calling into question the necessity of their deaths.

Envisioning the Future

Ideas about a postshuttle space program proliferated even during the shuttle's development. Concerned that piecemeal improvements would not fully address the shuttle's vulnerabilities, NASA and its contractors returned to the drawing board even before it flew, imagining more sophisticated architectures that might follow. As a result, instead of fixing the shuttle, NASA first spent thirty-five years attempting to replace it with a spaceplane far more complex and unworkable—fully reusable single-stage-to-orbit (SSTO) spaceplanes that previously had been rejected by the Air Force and NASA as fantastical. Pushing against the physical limits of chemical propulsion, these remarkable designs would have approached the capabilities of the space vehicles in science fiction: lifting off the ground and into space and returning without shedding engines and empty fuel tanks as previous craft, including the space shuttle, had. Building such craft proved as vexing as their theoretical sophistication suggested—a space vehicle light enough to lift all the fuel it would need into space would offer precious little mass allowance for anything else, providing the Pyrrhic victory of a spacecraft that arrived in orbit, only to do nothing at all.

It was only the end of the space shuttle program in 2011 that finally forced NASA to confront the Faustian bargain that it had made with President Nixon to approve it in 1972. Now reactive rather than proactive, the agency returned to capsules and expendable boosters for human spaceflight needs, a separation of military, scientific, and commercial spaceflight operations; as well as the complete rejection of private space tourism, which seemed foolhardy in light of spaceflight's inherent danger. Unable to sustain funding commitments and political interest in any of the shuttle's exotic replacements—from the NASP to the X-33 and the X-38—NASA ultimately retreated partially into its past, in the form of the shuttle-derived Space Launch System (SLS) and the Apollo-style Orion spacecraft.

At this point, a small number of private corporations, some funded by wealthy space enthusiasts, jumped into the human spaceflight business, in many cases repurposing existing NASA technology to create launch vehicles and space capsules offering reusable components and lower manufacturing and operating costs. Although members of the public were apt to view their arrival as proof of the imminent privatization of spaceflight, the new manufacturers (run by entrepreneurs who had prospered in other industries) sought principally to compete for the same government contracts to launch military and civilian satellites previously monopolized by large defense conglomerates, using technology developed through massive public investment in space technology. If anything, the movement toward so-called privatization was a devolution rather than an evolution of spaceflight, away from complex, massively funded flight vehicles and toward smaller, less capable ones built to inexpensively launch spy satellites and service the ISS in lieu of the space shuttle.

Although interest in winged space vehicles remained after the end of the shuttle program, the unique set of circumstance that produced the space shuttle of 1972 are not likely to rise again anytime soon: while the specter of international competition continues in the early twenty-first century, wariness of spaceflight's expense has slowed the pace of deep space exploration, and the rise of populist and authoritarian governments worldwide has elevated military space programs, undermined public faith in scientific expertise, and spurred a retreat from the kind of global cooperation the shuttle was created, in part, to foster. There appears to be little call now for dramatic, government-funded spaceflight outside of political bluster and hollow, unfunded calls to reclaim past glories with a return to the Moon.[27] NASA's space shuttles, meanwhile, have long since transitioned from technical infrastructures to museum exhibits. Although the orbiters are still serving the US as public display objects in New York, Florida, Washington, D.C., and Los Angeles, they do so principally as relics of a more optimistic, though not necessarily wiser, time.

Coda

Given NASA's exaggerated expectations for the space shuttle, it is not surprising that the vehicle failed to live up to them, and in that failure, the shuttle tells us more about the vulnerabilities of all big technologies than the merits of any one machine. The shuttle's problems are, in short, common to many technologies that fail. For example, it was a widely held perception of the spaceplane's inevitability that led to its creation, but historians of technology have long been wary of notions of technological fate. Building the shuttle required particular individuals and agencies to make discrete choices, and many of these were made poorly. Mechanisms designed to evaluate the merits of shuttle concepts collapsed under the weight of institutional pressure and political corruption. Had these mechanisms worked, the TAOS shuttle might have been rejected; its existence was not inexorable. This is a lesson that everyone should keep in mind as new generations of enthusiasts demand that people permit new technologies to become ubiquitous because the tide of history somehow demands it.

Another critical lesson offered by the space shuttle is that the responsibility to design ethically implies an obligation to do so from the very beginning. The shuttle was a knowingly flawed solution to a problem, later billed as a success, but no amount of postconstruction remediation could fix its flaws. Post-hoc solutions to design problems are often doomed to failure because fixing a broken technology often entails corrections more expensive than building something properly the first time. Creating elaborate procedures to suggest that a dangerous infrastructure is safe are even more problematic; while they may assuage the guilt of the operators of dangerous technologies, they seldom remove or redeem their vulnerabilities. The question of whether engineers have obligations to the public has challenged them for a century, with no clear answer emerging, but in the stories of Challenger and Columbia, events overtook theory.[28] The consequence of unethical design is a loss of faith in the craft of engineering itself, and, too frequently, loss of life.

Meanwhile, the problem with explanations of failure rooted in organizational dynamics or system complexity is not that these explanations are never wrong, but that when they are, they obscure responsibility, conceal agency, and make tragedies appear inevitable. While NASA's management of the Challenger and Columbia disasters earned it little praise, the vehicle was not too complex to manage; it was merely worse than many safer alternatives. The management deficiencies that Vaughan noted were doubtless present, but they are an insufficient explanation for the shuttle's woes. The lesson of the shuttle is not the importance of overseeing complex technological programs well (this should be obvious), but rather that work cultures should never be blamed for bad design or the malfeasance of individuals, no matter how comforting it might be to do so.

Finally, the shuttle highlights two critical issues in modern society's willingness to tolerate risk. The Challenger disaster highlights the discrepancy between NASA insiders' understanding of risk and the degree of risk that they communicated to the public because they did not trust the public to make proper decisions otherwise. Indeed, Richard Feynman found the Challenger's 1/100 risk of catastrophic failure appropriate; he merely objected to NASA's lack of disclosure of it before Challenger flew with a passenger who could not appreciate its dangers. The hype surrounding the shuttle was ultimately so deafening that honest and accurate assessments of the shuttle's vulnerabilities failed to reach those chosen to fly it. The case of Columbia also highlights the dangers of "decision-making by statistics," as Sean O'Keefe, NASA's administrator and self-proclaimed chief "bean counter," forbade a spacewalk to repair the stricken shuttle in orbit due to calculations indicating high risk and low probability of success. To astronauts willing to risk anything to save their comrades, this judgment was cowardly and heartless, the opposite of the ethos that they had hoped to instill in NASA.

Dark Star is a case study, thus, not merely in the social construction of technological systems or the path dependence of spaceflight technologies, but in the planned obsolescence of the high-tech world, the importance of ethical design, and the role of human agency in the management of risk. For NASA, this story offers both good and bad news. The good news is that the broad indictment of the agency's management culture during the shuttle years is probably unfair: analyzed in a vacuum, every large organization is deficient, and the shuttle's glaring problems would likely have damaged the vehicles even if NASA had been more cautious. In the case of Columbia, more evidence exists for the agency's poor management of a contingency after it happened, but these actions can't be blamed for the failure that precipitated it. The worse news is that the nation gambled its astronauts on a craft built halfway between a bet and a dare: a risky placeholder for a better future that NASA could glimpse, but never grasp.

Notes

Introduction

1. *Dark Star* (Bryanston Distributing, 1974).

2. *2001: A Space Odyssey* (Metro-Goldwyn-Mayer, 1968).

3. The ill-fated name "DarkStar" was eventually assigned to a 1990s joint Lockheed, Boeing, and Defense Advanced Research Projects Agency (DARPA) project to develop a reconnaissance drone with low radar observability. The first prototype crashed in 1996, and the program was canceled three years later. National Air and Space Museum (NASM), "Lockheed Martin/Boeing RQ-3A DarkStar," accessed June 5, 2020, https://airandspace.si.edu/collection-objects/lockheed-martin-boeing-rq-3a-darkstar/nasm_A20070230000.

4. For example, see Roger D. Launius and Howard E. McCurdy, eds., *Spaceflight and the Myth of Presidential Leadership* (Urbana: University of Illinois Press, 1997).

5. Walter A. McDougall, *. . . the Heavens and the Earth: A Political History of the Space Age* (Baltimore: Johns Hopkins University Press, 1997), 141.

6. For example, see Alexander C. T. Geppert, ed., *Imagining Outer Space: European Astroculture in the Twentieth Century*, 2d ed. (London: Palgrave Macmillan, 2018); Howard E. McCurdy, *Space and the American Imagination* (Washington, DC: Smithsonian Institution Press, 1997).

7. For example, see, generally, Neil M. Maher, *Apollo in the Age of Aquarius* (Cambridge, MA: Harvard University Press, 2017); and Patrick McCray, *The Visioneers: How a Group of Elite Scientists Pursued Space Colonies, Nanotechnologies, and a Limitless Future*, Course Book (Princeton, NJ: Princeton University Press, 2013).

8. For example, see Gerard K. O'Neill, *High Frontier: Human Colonies in Space* (New York: William Morrow and Company, 1977).

9. National Aeronautics and Space Administration (NASA), *Space Settlements: A Design Study*, SP-413 (Washington, DC: NASA, 1977).

10. *Space Is the Place* (North American Star System, 1974).

11. Gerard J. Degroot, *Dark Side of the Moon: The Magnificent Madness of the American Lunar Quest* (New York: New York University Press, 2006), 93.

12. James B. Gilbert, *Redeeming Culture: American Religion in an Age of Science* (Chicago: University of Chicago Press, 1997), 243.

13. Stewart Brand, ed., *The Whole Earth Catalog: Access to Tools* (Menlo Park, CA: Portola Institute, 1968).

14. Contra, McDougall, . . . *the Heavens and the Earth*, 421.

15. See, for example, Robert W. Smith, *The Space Telescope: A Study of NASA, Science, Technology, and Politics* (New York: Cambridge University Press, 1989).

16. Andrew J. Butrica, *Single Stage to Orbit: Politics, Space Technology, and the Quest for Reusable Rocketry* (Baltimore: Johns Hopkins University Press, 2003), 34.

17. Richard M. Nixon, "The Statement by President Nixon, 5 January 1972," 1972, http://history.nasa.gov/stsnixon.htm.

18. Nixon, "Statement by President Nixon"

19. See, for example, Roger D. Launius, "The Strange Career of the American Spaceplane: The Long History of Wings and Wheels in Human Space Operations," *Centaurus* 55, no. 4 (November 1, 2013): 412–432. See also Amy Paige Kaminski, *Sharing the Shuttle with America: NASA and Public Engagement after Apollo* (PhD diss., Science and Technology in Society, Blacksburg: Virginia Polytechnic Institute, 2015).

20. Butrica, *Single Stage to Orbit*, 7.

21. John M. Logsdon, *After Apollo? Richard Nixon and the American Space Program* (Palgrave MacMillan, 2015), 45.

22. Caspar W. Weinberger and George P. Shultz, "Memorandum for the President, 'Future of NASA,' August 12, 1971," in *Exploring the Unknown: Selected Documents in the History of the U.S. Civil Space Program*, ed. John M. Logsdon (Washington, DC: National Aeronautics and Space Administration, 1995), 546.

23. McDougall, . . . *the Heavens and the Earth*, 423.

24. McDougall, . . . *the Heavens and the Earth*, 421.

25. See, for example, Alex Roland, "Triumph or Turkey?" *Discover*, 1985.

26. See, for example, Dennis R. Jenkins, *Space Shuttle: Developing an Icon, 1972–2013*, vol. 1 (Forest Lake: Specialty Press, 2016); Dennis R. Jenkins, *Space Shuttle: Developing an Icon, 1972–2013*, vol. 2 (Forest Lake: Specialty Press, 2016); Dennis R. Jenkins, *Space Shuttle: The History of the National Space Transportation System* (Cape Canaveral, FL: D. R. Jenkins, 2010); Valerie Neal, *Spaceflight in the Shuttle Era and beyond: Redefining*

Humanity's Purpose in Space, 2017; Roger D. Launius, John Krige, James I. Craig, and Ned Allen, *Space Shuttle Legacy: How We Did It and What We Learned* (Reston, VA: American Institute of Aeronautics and Astronautics, 2013).

27. See David Edgerton, *The Shock of the Old: Technology and Global History since 1900* (New York: Oxford University Press, 2007), xvii.

28. Edgerton, *The Shock of the Old*, 2, 137.

29. Kaminski, *Sharing the Shuttle with America*; David Meerman Scott and Richard Jurek, *Marketing the Moon: The Selling of the Apollo Lunar Program* (Cambridge, MA: MIT Press, 2014).

30. Lewis Mumford, *Pentagon of Power: The Myth of the Machine, Vol. II* (New York: Harcourt Brace Jovanovich, 1974).

31. For example, see Richard R. John, "Debating New Media: Rewriting Communications History," *Technology and Culture* 64, no. 2 (2023): 308–58; Gabrielle Hecht and Michel Callon, *The Radiance of France: Nuclear Power and National Identity after World War II* (Cambridge, MA: MIT Press, 2009); Thomas P. Hughes, *Networks of Power: Electrification in Western Society, 1880–1930* (Baltimore: Johns Hopkins University Press, 1983); Peter Fritzsche, *A Nation of Fliers: German Aviation and the Popular Imagination* (Cambridge, MA: Harvard University Press, 1992); Roger E. Bilstein, *Flight in America, 1900–1983: From the Wrights to the Astronauts* (Baltimore: Johns Hopkins University Press, 1984); Daniel R. Headrick, *The Tools of Empire: Technology and European Imperialism in the Nineteenth Century* (New York: Oxford University Press, 1981).

32. Robert C. Seamans, Jr., and Frederick I. Ordway III, "Lessons of Apollo for Large-Scale Technology," in *Between Sputnik and the Shuttle: New Perspectives on American Astronautics*, ed. Frederick C. Durant, AAS History Series v. 3 (San Diego: Published for American Astronautical Society by Univelt, 1981), 269.

33. For example, see Ruth Schwartz Cowan and Matthew H. Hersch, *A Social History of American Technology*, 2d ed. (New York: Oxford University Press, 2017); Martin Campbell-Kelly and William Aspray, *Computer: A History of the Information Machine* (Boulder, CO: Westview Press, 2004).

34. For example, see Roger E. Bilstein, *Stages to Saturn: A Technological History of the Apollo/Saturn Launch Vehicles* (Washington, DC: Scientific and Technical Information Branch, NASA, 1980); Barton C. Hacker and James M. Grimwood, *On the Shoulders of Titans: A History of Project Gemini* (Washington, DC: NASA, 1978); Loyd S. Swenson, Jr., James M. Grimwood, and Charles C. Alexander, *This New Ocean: A History of Project Mercury* (Washington, DC: Scientific and Technical Information Division, Office of Technology Utilization, NASA, 1966); C. M. Green and M. Lomask, *Vanguard: A History* (Washington, DC: Smithsonian Institution Press, 1971).

35. For example, see Smith, *The Space Telescope*; P. E. Mack, *Viewing the Earth: The Social Construction of the Landsat Satellite System* (Cambridge, MA: MIT Press, 1990).

36. For example, see Roger D. Launius and Aaron K. Gillette, eds., *Toward a History of the Space Shuttle: An Annotated Bibliography* (Washington, DC: NASA History Office, 1992); Malinda K. Goodrich, Alice R. Buchalter, and Patrick M. Miller, eds., *Toward a History of the Space Shuttle: An Annotated Bibliography, Part 2 (1992–2011)* (Washington, DC: NASA History Program Office, 2012). See also Roger D. Launius, "Summer Reading: Indispensable Books on the History of the Space Shuttle," *Roger Launius's Blog*, June 12, 2017, https://launiusr.wordpress.com/2017/06/12/a-shelf-of-indispensable-books-on-the-space-shuttle/; Roger D. Launius, "A Baker's Dozen of Key Historical Books about the Space Shuttle," *Roger Launius's Blog*, October 19, 2011, https://launiusr.wordpress.com/2011/10/19/a-baker%e2%80%99s-dozen-of-key-historical-books-about-the-space-shuttle/.

37. For example, see Judy A. Rumerman, *U.S. Human Spaceflight: A Record of Achievement, 1961–2006* (Washington, DC: NASA, 2007).

38. T. A. Heppenheimer, *The Space Shuttle Decision, 1965–1972* (Washington, DC: Smithsonian Institution Press, 2002); T. A. Heppenheimer, *Development of the Shuttle, 1972–1981* (Washington, DC: Smithsonian Institution Press, 2002).

39. Wayne Hale, ed., *Wings in Orbit: Scientific and Engineering Legacies of the Space Shuttle, 1971–2010* (Washington, DC: NASA, 2010).

40. John M. Logsdon, ed., *Exploring the Unknown: Selected Documents in the History of the U.S. Civil Space Program*, vol. 1 (Washington, DC: NASA, 1995).

41. Jenkins, *Space Shuttle: The History of the National Space Transportation System*, 2010; Jenkins, *Space Shuttle: Developing an Icon, 1972–2013*, 3 vols., 2016.

42. Launius et al., *Space Shuttle Legacy*.

43. For example, see Hans Mark, *The Space Station: A Personal Journey* (Durham, NC: Duke University Press, 1987).

44. For example, see Donald K. Slayton and Michael Cassutt, *Deke! U.S. Manned Space: From Mercury to the Shuttle* (New York: St. Martin's Press, 1994); Christopher C. Kraft Jr., *Flight: My Life in Mission Control* (New York: Dutton, 2001).

45. Thomas D. Jones, *Sky Walking: An Astronaut's Memoir* (New York: Smithsonian Books, 2006).

46. Tom Wolfe, *The Right Stuff* (New York: Farrar, Straus, and Giroux, 1979).

47. Mike Mullane, *Riding Rockets: The Outrageous Tales of a Space Shuttle Astronaut* (New York: Scribner, 2006).

48. For example, see Homer Hickam, "What Makes an Astronaut Crack?" *Los Angeles Times*, February 9, 2007, VCA25; *Lucy in the Sky* (Fox Searchlight Pictures, 2019).

49. David Hitt and Heather R. Smith, *Bold They Rise: The Space Shuttle Early Years, 1972–1986* (Lincoln: University of Nebraska Press, 2014); Rick Houston, *Wheels Stop:*

The Tragedies and Triumphs of the Space Shuttle Program, 1986–2011 (Lincoln: University of Nebraska Press, 2013).

50. Smith, *The Space Telescope.*

51. Joseph D. Atkinson and Jay M. Shafritz, *The Real Stuff: A History of NASA's Astronaut Recruitment Program* (New York: Praeger, 1985); Howard E. McCurdy, *Inside NASA: High Technology and Organizational Change in the U.S. Space Program* (Baltimore: Johns Hopkins University Press, 1993).

52. Bryan Burrough, *Dragonfly: NASA and the Crisis aboard the MIR* (New York: HarperCollins, 1998).

53. Stephen J. Garber, *Birds of a Feather? How Politics and Culture Affected the Designs of the U.S. Space Shuttle and the Soviet Buran* (master's thesis, Virginia Tech, Blacksburg, 2002); Stephen J. Garber, "Why Does the Space Shuttle Have Wings? A Look at the Social Construction of Technology in Air and Space," *The Flight of STS-1*, April 5, 2001, https://history.nasa.gov/sts1/pages/scot.html; Launius, "The Strange Career of the American Spaceplane."

54. Logsdon, *After Apollo?*

55. Bettyann H. Kevles, *Almost Heaven: The Story of Women in Space* (New York: Basic Books, 2003).

56. Amy E. Foster, *Integrating Women into the Astronaut Corps: Politics and Logistics at NASA, 1972–2004* (Baltimore: Johns Hopkins University Press, 2011).

57. For example, see James E. David, *Spies and Shuttles: NASA's Secret Relationships with the DOD and CIA* (Gainesville: University Press of Florida, 2015).

58. See, for example, John M. Logsdon, *Ronald Reagan and the Space Frontier*, Palgrave Studies in the History of Science and Technology (Cham, Switzerland: Springer International Publishing AG, 2018); Logsdon, *After Apollo?*

59. Recently, several scholars have returned to Project Apollo to assess the lunar landings as cultural phenomena connected to dramatic cultural and social shifts in the US. Neil Maher's recent *Apollo in the Age of Aquarius* connects Apollo to the rise of American counterculture, while Kendrick Oliver's *To Touch the Face of God: The Sacred, the Profane, and the American Space Program, 1957–1975* offers a thoughtful mediation on the religious implications of Apollo. Maher, *Apollo in the Age of Aquarius*; Kendrick Oliver, *To Touch the Face of God: The Sacred, the Profane and the American Space Program, 1957–1975* (Baltimore: Johns Hopkins University Press, 2013). And Matthew Tribbe's *No Requiem for the Space Age* offers a philosophical take on Apollo's abrupt termination and the retreat of American technological utopianism. Matthew D. Tribbe, *No Requiem for the Space Age: The Apollo Moon Landings and American Culture* (Oxford: Oxford University Press, 2014).

60. Neal, *Spaceflight in the Shuttle Era and Beyond.*

61. Margaret A. Weitekamp, *Space Craze: America's Enduring Fascination with Real and Imagined Spaceflight* (Washington, DC: Smithsonian Books, 2022).

62. For example, see Alexander C. T. Geppert, ed., *Limiting Outer Space: Astroculture after Apollo* (Palgrave Macmillan UK, 2018); Neal, *Spaceflight in the Shuttle Era and Beyond*; Kaminski, *Sharing the Shuttle with America*.

63. Jenkins, *Space Shuttle*, 2016.

64. McCurdy, *Inside NASA*.

65. For example, see Neil Schlager, *When Technology Fails: Significant Technological Disasters, Accidents, and Failures of the Twentieth Century* (Detroit: Gale Research, 1994). See also William B. Rouse, *Failure Management: Malfunctions of Technologies, Organizations, and Society* (Oxford: Oxford University Press, 2021); Jonathan Coopersmith, *Faxed: The Rise and Fall of the Fax Machine*, Johns Hopkins Studies in the History of Technology (Baltimore: Johns Hopkins University Press, 2015); Norbert J. Delatte, *Beyond Failure: Forensic Case Studies for Civil Engineers* (Reston, VA: ASCE Press, 2009).

66. Scott A. Sandage, *Born Losers: A History of Failure in America* (Cambridge, MA: Harvard University Press, 2005), 11.

67. Edward Jones-Imhotep, *The Unreliable Nation: Hostile Nature and Technological Failure in the Cold War* (Cambridge, MA: MIT Press, 2017), 10–13.

68. Roger D. Launius, "Assessing the Legacy of the Space Shuttle," *Space Policy* 22 (2006): 226–234.

69. Matthew H. Hersch, "Using the Shuttle: Operations in Orbit," in *Space Shuttle Legacy: How We Did It and What We Learned*, ed. Roger D. Launius, John Krige, James I. Craig, and Ned Allen, 191–214, Library of Flight Series (Reston, VA: American Institute of Aeronautics and Astronautics, Inc, 2013).

70. Harry M. Collins and Trevor J. Pinch, *The Golem at Large: What You Should Know about Technology* (New York: Cambridge University Press, 1998), 30–56.

71. Roland, "Triumph or Turkey?"

72. Diane Vaughan, *The Challenger Launch Decision: Risky Technology, Culture, and Deviance at NASA* (Chicago: University of Chicago Press, 1996), xv.

73. Vaughan, *The Challenger Launch Decision*. See also Claus Jensen, *No Downlink: A Dramatic Narrative about the Challenger Accident and Our Time* (New York: Farrar, Straus & Giroux, 1996); Allan J. McDonald, *Truth, Lies, and O-Rings: Inside the Space Shuttle Challenger Disaster* (Gainesville: University Press of Florida, 2009); Kevin Cook, *The Burning Blue: The Untold Story of Christa McAuliffe and NASA's Challenger Disaster* (New York: Henry Holt and Company, 2021).

74. For example, see Michael Cabbage, *Comm Check . . . : The Final Flight of Shuttle Columbia* (New York: Free Press, 2004); Philip Chien, *Columbia, Final Voyage: The Last Flight of NASA's First Space Shuttle* (New York: Copernicus Books, 2006).

75. For example, see Launius, "Assessing the Legacy of the Space Shuttle"; Roger D. Launius, "The Space Shuttle—Twenty-Five Years On: What Does It Mean to Have Reusable Access to Space?" *Quest* 13 (2006): 10; John M. Logsdon, "Was the Space Shuttle a Mistake?" *MIT Technology Review*, July 6, 2011, https://www.technologyreview.com/s/424586/was-the-space-shuttle-a-mistake/.

76. R. Jeffrey Smith, "Estrangement on the Launch Pad," *Science*, June 29, 1984, 1408.

77. Arwen Mohun, *Risk: Negotiating Safety in American Society* (Baltimore: Johns Hopkins University Press, 2013).

78. Edgerton, *The Shock of the Old.*

79. Mohun, *Risk*, 3.

80. Mohun, 5.

81. R. Nader, *Unsafe at Any Speed, the Designed-in Dangers of the American Automobile* (New York: Grossman, 1965).

82. "Deep Space Homer," *The Simpsons* (Fox, February 24, 1994).

83. Edgerton, *The Shock of the Old.*

84. Lee Vinsel and Andrew L. Russell, *The Innovation Delusion: How Our Obsession with the New Has Disrupted the Work That Matters Most* (New York: Currency, 2020).

85. Vance Packard, *The Waste Makers* (New York: D. McKay Co., 1960).

86. Joseph J. Corn, *The Winged Gospel: America's Romance with Aviation, 1900–1950* (New York: Oxford University Press, 1983).

87. For example, see Heppenheimer, *The Space Shuttle Decision, 1965–1972*; Launius, "The Strange Career of the American Spaceplane"; Garber, "Why Does the Space Shuttle Have Wings?"

88. Vaughan, *The Challenger Launch Decision.*

89. Charles Perrow, *Normal Accidents: Living with High-Risk Technologies: With a New Afterword and a Postscript on the Y2K Problem* (Princeton, NJ: Princeton University Press, 1999).

90. Rogers Commission, *Report of the Presidential Commission on the Space Shuttle Challenger Accident*, vol. 1 (Washington, DC: Government Printing Office, 1986).

91. Thomas P. Hughes, "Technological Momentum," in *Does Technology Drive History? The Dilemma of Technological Determinism*, ed. Merritt Roe Smith and Leo Marx (Cambridge, MA: MIT Press, 1994), 101–113.

1 The Silver Bird Comes to America, 1941–1963

1. E. Sänger and J. Bredt, *Tiberelnen Raketenantrleb for Fernboober (A Rocket Drive for Long-Range Bombers)*, trans. M. Hamermesh (Technical Information Branch, BUAER, Navy Department, 1952), 161; Irene Sänger-Bredt, "The Silver Bird Story," *Spaceflight* 15 (May 1973): 172.

2. Sänger and Bredt, *Tiberelnen Raketenantrleb for Fernboober(A Rocket Drive for Long-Range Bombers)*, 6; Irene Sänger-Bredt, "The Silver Bird Story."

3. The craft's 6,600 miles per hour top speed would be only about a third of that required to sustain a stable orbit. Roger D. Launius and Dennis R. Jenkins, *Coming Home: Reentry and Recovery from Space* (Washington, DC: Government Printing Office, 2012), 7.

4. For example, see Willy Ley, *Rockets, Missiles, and Space Travel*, Rev. and enl. ed (New York: Viking Press, 1957), 431–434.

5. For example, see Sänger and Bredt, *Tiberelnen Raketenantrleb for Fernboober(A Rocket Drive for Long-Range Bombers)*, 6, 50, 84, 85.

6. Eugen Sänger, *Rocket Flight Engineering*, NASA Technical Translations, F-223 (Washington, DC: NASA, 1965), 313.

7. Sänger and Bredt, *Tiberelnen Raketenantrleb for Fernboober (A Rocket Drive for Long-Range Bombers)*, 87.

8. R. Dale Reed and Darlene Lister, *Wingless Flight: The Lifting Body Story* (Washington, DC: NASA History Office, Office of Policy and Plans, 1997), 111 et seq.

9. George Raudzens, "War-Winning Weapons: The Measurement of Technological Determinism in Military History," *Journal of Military History* 54, no. 4 (1990): 425 et seq., https://doi.org/10.2307/1986064.

10. See Michael S. Sherry, *The Rise of American Air Power the Creation of Armageddon* (Philadelphia: University of Pennsylvania Press, 2006); Roger E. Bilstein, *Flight in America, 1900–1983: From the Wrights to the Astronauts* (Baltimore: Johns Hopkins University Press, 1984).

11. Wernher von Braun, F. I. Ordway, and D. Dooling, *Space Travel: A History*, 4th ed. (New York: Harper & Row, 1985).

12. See, generally, von Braun et al., *Space Travel*.

13. See, generally, Milton Lehman, *This High Man: The Life of Robert H. Goddard* (New York: Farrar, 1963).

14. Roger D. Launius, "The Strange Career of the American Spaceplane: The Long History of Wings and Wheels in Human Space Operations," *Centaurus* 55, no. 4 (November 1, 2013): 414.

15. Frank J. Malina, "Interview with Frank J. Malina," Oral History, 1980, http://resolver.caltech.edu/CaltechOH:OH_Malina_F; Erik M. Conway, "From Rockets to Spacecraft: Making JPL a Place for Planetary Science," *Engineering and Science* 70, no. 4 (2007): 2–10; Peter J. Westwick, *Into the Black: JPL and the American Space Program, 1976–2004* (New Haven, CT: Yale University Press, 2011).

16. Frank J. Malina, "The Rocket Pioneers," *Engineering & Science*, November 1986, 12.

17. The article draws from the author's research and previous published work, including Cassandra Steer and Matthew H. Hersch (eds.), *War and Peace in Outer Space: Law, Policy, and Ethics* (Oxford: Oxford University Press, 2021).

18. See, generally, Westwick, *Into the Black: JPL and the American Space Program, 1976–2004*.

19. T. A. Heppenheimer, *The Space Shuttle Decision, 1965–1972* (Washington, DC: Smithsonian Institution Press, 2002), 43.

20. See, for example, George Pendle, *Strange Angel: The Otherworldly Life of Rocket Scientist John Whiteside Parsons* (New York: Houghton Mifflin Harcourt, 2006).

21. Malina, "Interview with Frank J. Malina," 14.

22. For example, see Frank J. Malina, *Notes on Tour of Inspection of the United Kingdom and France*" (Pasadena: California Institute of Technology, 1944).

23. Sänger-Bredt, "The Silver Bird Story," 170.

24. Sänger-Bredt, "The Silver Bird Story," 166.

25. Sänger, *Rocket Flight Engineering*; Sänger-Bredt, "The Silver Bird Story," 171.

26. Maurice J. Zucrow, *Principles of Jet Propulsion and Gas Turbines* (New York: John Wiley, 1948), 465.

27. See, generally, Sänger-Bredt, "The Silver Bird Story," 169.

28. Sänger-Bredt, "The Silver Bird Story," 167.

29. Sänger-Bredt, "The Silver Bird Story," 170.

30. Sänger-Bredt, "The Silver Bird Story," 173.

31. Sänger-Bredt, "The Silver Bird Story," 172–175.

32. Launius and Jenkins, *Coming Home*, 11. With Germany unable to develop an atomic bomb during World War II, both rocket weapons remained impractical through V-E Day, their destructive power being insufficient to compensate for their targeting inaccuracies.

33. For example, see Michael J. Neufeld, *Spaceflight: A Concise History* (Cambridge, MA: MIT Press, 2018), 19–20.

34. Michael Neufeld, "Wernher von Braun, the SS, and Concentration Camp Labor: Questions of Moral, Political, and Criminal Responsibility," *German Studies Review*, 2002.

35. Michael J. Neufeld, *Von Braun: Dreamer of Space, Engineer of War* (New York: A. A. Knopf, 2007), 101.

36. Dennis R. Jenkins, *Space Shuttle: Developing an Icon, 1972–2013*, Vol. 1, 2016, 1: 6.

37. Krafft A. Ehricke, "Peenemuende Rocket Center," January 3, 1950, 37, Box 1, Folder 4, Krafft Arnold Ehricke Papers, National Air and Space Museum Archives, Washington, DC.

38. Ley, *Rockets, Missiles, and Space Travel.*

39. Stephen J. Garber, "Why Does the Space Shuttle Have Wings? A Look at the Social Construction of Technology in Air and Space (IV. SCOT and the Shuttle's Wings)," July 3, 2014, https://history.nasa.gov/sts1/pages/scotc.html.

40. For example, see Charles J. Bauer, "'Rocket' Interceptor Tested," *New York Times*, January 15, 1939, XX8.

41. Thomas Pynchon, *Gravity's Rainbow* (New York: Bantam Books, 1974).

42. Sänger-Bredt, "The Silver Bird Story," 172.

43. For example Sänger and Bredt, *Tiberelnen Raketenantrleb for Fernboober (A Rocket Drive for Long-Range Bombers).*

44. Joel Carpenter, "PROJECT 1947: German Hypersonic Bomber Prototype?" November 11, 2011, http://www.project1947.com/gfb/antiplofer.htm.

45. Sänger-Bredt, "The Silver Bird Story," 170.

46. "Says Pilots Like Rocket Plane," *New York Times*, January 25, 1944, 4.

47. For example, see Zucrow, *Principles of Jet Propulsion and Gas Turbines.*

48. "Rocket Plane Downs Japanese Fighter," *New York Times*, May 18, 1945, 4.

49. *Rocky Jones, Space Ranger* (Roland Reed TV Productions, 1954).

50. "The Rocket Plane Is Here," *New York Times*, January 8, 1944, 12.

51. For example, see Neufeld, *Spaceflight*, 20.

52. For example, see Conway, "From Rockets to Spacecraft"; Malina, "The Rocket Pioneers."

53. "Rocket Plane Held Army's Prime Need," *New York Times*, March 8, 1946, 8.

54. Edward W. Constant, *The Origins of the Turbojet Revolution* (Baltimore: Johns Hopkins University Press, 1980).

55. Barton C. Hacker, "Whoever Heard of Nuclear Ramjets? Project Pluto, 1957–1964," *Journal of the International Committee for the History of Technology* 1 (1995): 85–98.

56. Alex Roland, "Twin Paradoxes of the Space Age," *Nature* 392, no. 6672 (March 1998): 143.

57. Jenkins, *Space Shuttle: Developing an Icon*, 2016, 1:10 et seq.

58. H. Allaway, *The Space Shuttle at Work* (Washington, DC: NASA, 1979), 30.

59. Launius, "The Strange Career of the American Spaceplane," 415–416; Darrell Romick, "Concept for a Manned Earth-Satellite Terminal Evolving from Earth-to-Orbit Ferry Rockets," September 17, 1956, Folder 007923, NASA Historical Reference Collection, NASA Headquarters, Washington, DC.

60. Werner von Braun, "Can We Get to Mars?" *Collier's*, 1954.

61. Werner von Braun, *The Mars Project* (Urbana: University of Illinois Press, 1953).

62. Neufeld, *Von Braun: Dreamer of Space, Engineer of War*, 241.

63. Clarence J. Geiger, *History of the X-20A Dyna-Soar*, vol. 1 (Wright-Patterson AFB: Historical Division, Aeronautical Systems Division Information Office, US Air Force, 1963), 9.

64. Launius, "The Strange Career of the American Spaceplane," 421.

65. Launius, "The Strange Career of the American Spaceplane," 421.

66. R. E. Bilstein and F. W. Anderson, *Orders of Magnitude: A History of the NACA and NASA, 1915–1990* (Washington, DC: NASA, Office of Management, Scientific and Technical Information Division, 1989).

67. "The upper stage of a ferry or orbital carrier, however, must have wings." For example, Krafft A. Ehricke, "Project Orbital Carrier (1st Edition)," May 1952, n.p., Box 1, Folder 6, Krafft Arnold Ehricke Papers, National Air and Space Museum Archives, Washington, DC.

68. Sänger-Bredt, "The Silver Bird Story," 167.

69. Sänger-Bredt, "The Silver Bird Story," 167.

70. Mark Wade, "Saenger," *Encyclopedia Astronautica*, March 30, 2019, http://www.astronautix.com/s/saenger.html.

71. Sänger-Bredt, "The Silver Bird Story," 175.

72. For example, see Neufeld, *Spaceflight*, 19.

73. Roger E. Bilstein, *Flight in America: From the Wrights to the Astronauts*, 3rd ed. (Baltimore: Johns Hopkins University Press, 2001).

74. Sänger-Bredt, "The Silver Bird Story," 176.

75. Sänger-Bredt, "The Silver Bird Story," 166.

76. Sänger-Bredt, "The Silver Bird Story," 166.

77. Walter Dornberger, *V-2* (New York: Viking Press, 1954), 2.

78. Dornberger, *V-2,* 251.

79. Dornberger, *V-2,* 227.

80. For example, "Hail Rocket Plane in First Test Hop: Army's New High-Speed Rocket Plane in Test Flight," *New York Times,* December 11, 1946,1; Dennis R. Jenkins, Tony Landis, and Jay Miller, *American X-Vehicles: An Inventory—X-1 to X-50,* Centennial of Flight (Washington, DC: NASA, 2003).

81. Willy Ley, *The Conquest of Space* (New York: Viking Press, 1949); see also Catherine L. Newell, *Destined for the Stars: Faith, the Future, and America's Final Frontier* (Pittsburgh: University of Pittsburgh Press, 2019), 138.

82. For example, see Geiger, *History of the X-20A Dyna-Soar,* 1:5.

83. Sänger-Bredt, "The Silver Bird Story," 172.

84. See, for example, Geiger, *History of the X-20A Dyna-Soar*; Clarence J. Geiger, *Strangled Infant: The Declassified History of the Dyna-Soar Space Plane* (St. Petersburg, FL: Red and Black Publishers, 2019).

85. Ley, *Rockets, Missiles, and Space Travel,* 382–383.

86. See, generally, Green and Lomask, *Vanguard: A History.*

87. "Soviet Rocket Plane Foreseen," *New York Times,* February 18, 1958, 17.

88. See, for example, Geiger, *History of the X-20A Dyna-Soar,* 1:4.

89. Andrew L. Jenks, *Collaboration in Space and the Search for Peace on Earth,* Anthem Series on Russian, East European and Eurasian Studies (London: Anthem Press, 2022), 20–21.

90. Asif A. Siddiqi, *Challenge to Apollo: The Soviet Union and the Space Race, 1945–1974* (Washington, DC: NASA, NASA History Division, Office of Policy and Plans, 2000), 52–53.

91. Ray A. Williamson, "Developing the Space Shuttle: Early Concepts of a Reusable Launch Vehicle," in *Accessing Space,* vol. IV, Exploring the Unknown: Selected Documents in the History of the U.S. Civil Space Program (Washington, DC: NASA History Division, 1999), 162–163.

92. Richard Witkin, "2 Teams Bidding on Space Glider: Air Force Likely to Narrow Dyna-Soar Race to Boeing and Bell-Martin Groups," *New York Times,* May 11, 1958, 41.

93. For example, see Mark Wade, "Apollo L-2C," *Encyclopedia Astronautica*, 2019, http://www.astronautix.com/a/apollol-2c.html.

94. "Dyna-Soar Is Renamed X-20 by the Air Force," *New York Times*, June 27, 1962, 5.

95. Geiger, *History of the X-20A Dyna-Soar*, 1:64.

96. Geiger, *History of the X-20A Dyna-Soar*, 1:108.

97. A code name rather than an acronym, Corona—like other reconnaissance satellite program names—is properly written in upper- and lowercase letters, but this usage is inconsistent even within the government agency that operated the vehicles. Center for the Study of National Reconnaissance Classics, *The CORONA Story* (Chantilly, VA: Center for the Study of National Reconnaissance, 2013).

98. Richard Paul and Steven Moss, *We Could Not Fail: The First African Americans in the Space Program* (Austin: University of Texas Press, 2015), 102.

99. Siddiqi, *Challenge to Apollo*, 234.

100. Geiger, *History of the X-20A Dyna-Soar*, 1:115.

101. Hanson W. Baldwin, "Dyna-Soar or Dinosaur? Another Military Space Vehicle Due to Become Extinct in Conflict of Ideas View of Air Force Two Forces at Work," *New York Times*, March 16, 1963, 4.

102. Geiger, *History of the X-20A Dyna-Soar*, 1:115–16.

103. For example, see Office of the Director of Defense Research & Engineering, "Letter to Robert C. Truax," September 2, 1966, 1–2, Box 1, Folder 4, Robert C. Truax Collection, National Air and Space Museum Archives, Washington, DC.

104. See, generally, Siddiqi, *Challenge to Apollo*, 599–607.

105. Krafft A. Ehricke, "Extraterrestrial Imperative," in *Proceedings* (ION National Space Meeting on Space Shuttle, Space Station, Nuclear Shuttle Navigation, George C. Marshall Space Flight Center: Institute of Navigation, Huntsville, AL, 1971), 139–150.

106. Ehricke, "Extraterrestrial Imperative," 144–145.

2 Star Clipper and Other Fantasies, 1963–1969

1. See, generally, George Raudzens, "War-Winning Weapons: The Measurement of Technological Determinism in Military History," *Journal of Military History* 54, no. 4 (1990): 403–434. https://doi.org/10.2307/1986064.

2. *Man in Space*, Ward Kimball (dir.) (Los Angeles: Walt Disney Productions, 1955); Michael J. Neufeld, *Von Braun: Dreamer of Space, Engineer of War* (New York: A. A.

Knopf, 2007); Wernher von Braun and Cornelius Ryan, *Conquest of the Moon* (New York: Viking Press, 1953); Werner von Braun, "Can We Get to Mars?" *Collier's*, 1954

3. *Conquest of Space,* Byron Haskins (dir.) (Los Angeles: Paramount Pictures, 1955).

4. Courtney Sheldon, "Planet Shuttle Proposed: Steps Outlined," *Christian Science Monitor*, December 9, 1959, 5.

5. Air Force Space Systems Division, "Lunar Expedition Plan: Lunex" (US Air Force Systems Command, May 1961); Robert Godwin, "The Forgotten Plans to Reach the Moon—Before Apollo," *Air & Space Magazine*, July 19, 2019, https://www.airspacemag.com/daily-planet/forgotten-plans-reach-moon-apollo-180972695/.

6. Space Task Group, "NASA Project A, Announcement 1," December 22, 1958, Folder 013880, NASA Historical Reference Collection, NASA Headquarters, Washington, DC.

7. M. P. Mackowski, *Testing the Limits: Aviation Medicine and the Origins of Manned Space Flight* (College Station: Texas A&M University Press, 2006), 168; D. Flickinger, "Biomedical Aspects of Space Flight," in *Man in Space: The United States Air Force Program for Developing the Spacecraft Crew*, ed. K. F. Gantz (New York: Duell, 1959), 46.

8. Richard Witkin, "X-15 Rocket Plane Is Unveiled by U.S.," *New York Times*, October 16, 1958, 1; Gladwin Hill, "X-15 Rocket Plane in Powered Flight," *New York Times*, September 18, 1959, 1.

9. For example, see Dennis R. Jenkins, *Hypersonics before the Shuttle: A Concise History of the X-15 Research Airplane* (National Aeronautics and Space Administration, NASA Office of Policy and Plans, NASA History Office, NASA Headquarters, 2000).

10. Michelle Evans, *The X-15 Rocket Plane: Flying the First Wings into Space*, (Lincoln: University of Nebraska Press, 2013), 412.

11. Marvin Miles, "Advanced X-15 Proposed for Manned Orbital Space Ship: Current Tests Designed for 50-Mile Altitude Because of Heat Problem Advanced X-15 Urged for Manned Space Ship," *Los Angeles Times*, January 11, 1960, B1.

12. See, generally, Chris Petty and Dennis R. Jenkins, *Beyond Blue Skies: The Rocket Plane Programs That Led to the Space Age*, ill. ed. (Lincoln: University of Nebraska Press, 2020).

13. Roger D. Launius, "The Strange Career of the American Spaceplane: The Long History of Wings and Wheels in Human Space Operations," *Centaurus* 55, no. 4 (November 1, 2013): 418.

14. For example, see Robert C. Truax, "From Airlines to Spacelines?" 1963, 9, Box 18, Folder 11, Robert C. Truax Collection, National Air and Space Museum Archives, Washington, DC; Robert C. Truax, "The Pressure-Fed Booster–Dark Horse of the Space Race" (XIXth International Astronautical Congress, 1968), 2, Box 18, Folder 12, Robert C. Truax Collection, National Air and Space Museum Archives, Washington, DC.

15. Matthew H. Hersch, *Inventing the American Astronaut*, Palgrave Studies in the History of Science and Technology (New York: Palgrave Macmillan, 2012), chap. 1.

16. Joseph D. Atkinson and Jay M. Shafritz, *The Real Stuff: A History of NASA's Astronaut Recruitment Program* (New York: Praeger, 1985), 32–33.

17. Colin Burgess, *Selecting the Mercury Seven: The Search for America's First Astronauts* (Chichester, UK: Springer-Praxis, 2011), 30.

18. For example, see "Seven Pilots Picked for Satellite Trips: 7 Chosen by U.S. as Space Pilots," *New York Times*, April 7, 1959, 1; "2 Women Seek Roles as U.S. Space Pilots," *New York Times*, July 7, 1963, 15; "Kennedy Phones Salute to Pilot: Watches Astronaut on TV," *New York Times*, July 22, 1961, 10; "Space Pilots Get Training on Jets: Future Astronaut Describes Simulated Re-entry," *New York Times*, November 10, 1963, 88.

19. See, generally, Hersch, *Inventing the American Astronaut*.

20. NASA Space Task Group, "Handwritten Revisions to 12/30/1960 Telex from Loudon Wainwright, Life Magazine," 1960, Records of the National Aeronautics and Space Administration. National Archives and Records Center, College Park, MD; see, generally, Matthew H. Hersch, "'Capsules Are Swallowed': The Mythology of the Pilot in American Spaceflight," in *Spacefarers: Images of Astronauts and Cosmonauts in the Heroic Era of Spaceflight*, ed. Michael J. Neufeld (Washington, DC: Smithsonian Institution, 2013).

21. "X-15 Test Pilot Decorated as First Winged Astronaut," *New York Times*, June 4, 1963, 9.

22. Howard E. McCurdy, *Space and the American Imagination* (Washington, DC: Smithsonian Institution Press, 1997), 88–91.

23. See, generally, Matthew H. Hersch, "Return of the Lost Spaceman: America's Astronauts in Popular Culture, 1959–2006," *Journal of Popular Culture* 44, no. 1 (February 1, 2011): 73–92, https://doi.org/10.1111/j.1540-5931.2010.00820.x; Matthew H. Hersch, "Redemptive Space: Duty, Death, and the Astronaut-Soldier, 1949–1969," in *We Are All Astronauts: The Image of the Space Traveler in Arts and Media*, ed. Henry Keazor (Berlin: Neofelis Verlag, 2019), 35–53.

24. Walter Cunningham, *The All-American Boys: An Inside Look at the American Space Program*, rev. ed. (New York: ibooks, 2003), 82–83.

25. Michael J. Neufeld, *Spacefarers: Images of Astronauts and Cosmonauts in the Heroic Era of Spaceflight* (Washington, DC: Smithsonian Institution, 2013); Michael Collins, *Carrying the Fire: An Astronaut's Journeys* (New York: Farrar, 1974).

26. "6 Pilots Assigned to Dyna-Soar Tests," *New York Times*, March 15, 1962, 30.

27. See, for example, David A. Mindell, *Digital Apollo: Human and Machine in Spaceflight* (Cambridge, MA: MIT Press, 2008).

28. See, for example, Barton C. Hacker, "The Gemini Paraglider: A Failure of Scheduled Innovation, 1961–64," *Social Studies of Science* 22, no. 2 (May 1, 1992): 387–406, https://doi.org/10.1177/030631292022002012.

29. See, for example, William C. Thompson and United States, *Dynamic Model Investigation of the Landing Characteristics of a Manned Spacecraft*, NASA TN D-2497 (Washington, DC: NASA Langley Research Center, 1965).

30. Hacker, "The Gemini Paraglider."

31. Hacker, "The Gemini Paraglider," 396.

32. Michael J. Neufeld, "The Von Braun Paradigm and NASA's Long-Term Planning for Human Spaceflight," in *NASA's First 50 Years: Historical Perspectives*, ed. Steven J. Dick (Washington, DC: NASA, 2010), 325–347.

33. H. H. Koelle, "Letter to Edward R. Diemer" (National Aeronautics and Space Administration, February 8, 1963), 1, Folder 008211, NASA Historical Reference Collection, NASA Headquarters, Washington, DC.

34. Society of Experimental Test Pilots, ed., "Testy Test Pilots Society (for Lack of a Better Name at This Point): Minutes of the First Organized Meeting," in *History of the First 20 Years* (Covina: Taylor Publishing, 1978), 2.

35. Ray A. Williamson, "Developing the Space Shuttle: Early Concepts of a Reusable Launch Vehicle," in *Accessing Space*, vol. IV, Exploring the Unknown: Selected Documents in the History of the U.S. Civil Space Program (Washington, DC: NASA History Division, 1999), 162.

36. Walter R. Dornberger, "The Recoverable, Reusable Space Shuttle," *Aeronautics and Astronautics*, November 1965, 88, Folder 007923, NASA Historical Reference Collection, NASA Headquarters, Washington, DC.

37. Dornberger, "Recoverable, Reusable Space Shuttle," 89.

38. For example, see "Lockheed Wins Spaceship Study Pact from NASA," *Boston Globe*, July 10, 1962, 11; "Lockheed Wins Contract to Design a Rocket Ship for a Space Shuttle," *Wall Street Journal*, July 10, 1962, 10.

39. Dennis R. Jenkins, *Space Shuttle: Developing an Icon, 1972–2013*, Vol. 1, 2016, 1:256.

40. Edward Z. Gray, "Memorandum for the Associate Administrator Re: Saturn S-IC Stage Recovery and Reusability Study," June 8, 1964, 1, Folder 008211, NASA Historical Reference Collection, NASA Headquarters, Washington, DC.

41. Donald H. Heaton, "Memorandum for the Files," August 23, 1960, Folder 008211, NASA Historical Reference Collection, NASA Headquarters, Washington, DC.

42. See, for example, T. A Heppenheimer, *Facing the Heat Barrier: A History of Hypersonics* (Washington, DC: NASA, History Division, 2006), 197 et seq.

43. Dornberger, "Recoverable, Reusable Space Shuttle," 91.

44. For example, see Truax, "Pressure-Fed Booster–Dark Horse of the Space Race."

45. For example, see Neufeld, *Von Braun: Dreamer of Space, Engineer of War*, 438.

46. Brian O'Leary, "The Space Shuttle: NASA's White Elephant in the Sky," *Bulletin of the Atomic Scientists* 29, no. 2 (February 1973): 37.

47. Logsdon, *After Apollo?* 62–63.

48. Jenkins, *Space Shuttle: Developing an Icon*, 1: 210.

49. George E. Mueller, "Honorary Fellowship Acceptance, British Interplanetary Society, University College, London, England," August 10, 1968, 16, https://historydms.hq.nasa.gov/MUELLER, GEORGE E.; "Next: Earth-Orbiting Vacations!" *Christian Science Monitor*, October 19, 1968.

50. J. Randy Taraborrelli, *The Hiltons: The True Story of an American Dynasty* (New York: Grand Central Publishing, 2014), 277.

51. Office of Manned Space Flight, "Review of Orbital Transportation Concepts-Low Cost Operations (Integrated Launch/Reentry Vehicle Systems; Round-Trip Operations)" (National Aeronautics and Space Administration, December 18, 1968), 17, Folder 008211, NASA Historical Reference Collection, NASA Headquarters, Washington, DC.

52. See, generally, Walter J. Boyne, *Beyond the Horizons: The Lockheed Story* (New York: Thomas Dunne Books, 1998).

53. Petty and Jenkins, *Beyond Blue Skies*, 165, 330.

54. See Andrew J. Butrica, *Single Stage to Orbit: Politics, Space Technology, and the Quest for Reusable Rocketry* (Baltimore: Johns Hopkins University Press, 2003), 91–93.

55. "Lockheed Wins Spaceship Study Pact from NASA."

56. For example, Lockheed Corporation, "Lockheed Annual Report: 1963" (Ann Arbor, MI: ProQuest Annual Reports, 1963), http://search.proquest.com/hnpnewyorktimes/docview/88197575/citation/64159BCECD54CE3PQ/6.

57. Boyne, *Beyond the Horizons*.

58. Neufeld, *Von Braun: Dreamer of Space, Engineer of War*, 438.

59. Office of Manned Space Flight, "Review of Orbital Transportation Concepts-Low Cost Operations," 15.

60. Jenkins, *Space Shuttle: Developing an Icon*, 1:164.

61. Jenkins, *Space Shuttle*, 2016, 1:183.

62. Jenkins, *Space Shuttle*, 2016, 1:182.

63. Mueller, "Honorary Fellowship Acceptance, British Interplanetary Society, University College, London, England," 17.

64. Logsdon, *After Apollo?* 159.

65. James C. Fletcher, "Memorandum to Peter M. Flanigan Re: New Names for the Shuttle," December 30, 1971, 1, Ms 202, Box 4, Folder 5, James Chipman Fletcher Papers. Special Collections and Archives. University of Utah, J. Willard Marriott. Salt Lake City.

66. James C. Fletcher, "Note to Herbert Rowe," August 12, 1976, 1, Ms 202, Box 42, Folder 4, James Chipman Fletcher Papers. Special Collections and Archives. University of Utah, J. Willard Marriott, Salt Lake City.

67. Herbert J. Rowe, "Note to Administrator," August 18, 1976, 1, Ms 202, Box 42, Folder 4, James Chipman Fletcher Papers. Special Collections and Archives. University of Utah, J. Willard Marriott, Salt Lake City.

68. Jenkins, *Space Shuttle: Developing an Icon*, 1:134.

69. Jenkins, *Space Shuttle: Developing an Icon*, 2016, 1:138–141.

70. For example, see "Model, Space Shuttle, North American Rockwell Expendable Booster Concept; 1:200," National Air and Space Museum, May 1, 2021, https://airandspace.si.edu/collection-objects/model-space-shuttle-north-american-rockwell-expendable-booster-concept-1200/nasm_A19760785000.

71. The section draws form the author's prior work, including Matthew H. Hersch, "Pathfinder to Profit: Lessons from the Space Shuttle Era" (NASA and the Rise of Commercial Space Symposium, NASA Marshall Space Flight Center History Office, 2021).

72. Gray, "Memorandum for the Associate Administrator," 1.

73. Gray, Memorandum for the Associate Administrator," 1.

74. Marvin Miles, "Wingless 'Lift' Vehicle Nearly Ready for Tests: Northrop Completing Craft That May Be Forerunner of Future Space Shuttle Ship," *Los Angeles Times*, April 23, 1965, 19; Evert Clark, "Rocket Plane May Let Astronauts Land at Airfields," *New York Times*, July 12, 1967, 1. See also R. Dale Reed and Darlene Lister, *Wingless Flight: The Lifting Body Story* (Washington, DC: NASA History Office, Office of Policy and Plans, 1997), 174.

75. Williamson, "Developing the Space Shuttle," 162.

76. Reed and Lister, *Wingless Flight*, 129 et seq.

77. James Oberg, "Did Pentagon Create Orbital Space Plane?" NBC News Space Analyst Special, MSNBC.com, March 6, 2006, http://www.nbcnews.com/id/11691989/ns/technology_and_science-space/t/did-pentagon-create-orbital-space-plane/.

78. See, for example, Neufeld, *Von Braun: Dreamer of Space, Engineer of War*, 450.

79. Darrell C. Romick, "METEOR, Jr.: A Preliminary Design Investigation of a Minimum Sized Ferry Rocket Vehicle of the METEOR Concept" (Goodyear Aircraft Corporation, n.d.), 1, Box 2, Darrell C. Romick Papers, National Air and Space Museum Archives, Washington, DC.

80. Jenkins, *Space Shuttle: Developing an Icon*, 1:204.

81. See, for example, Neufeld, *Von Braun: Dreamer of Space, Engineer of War*, 451.

82. Clark, "Rocket Plane May Let Astronauts Land at Airfields."

83. Jenkins, *Space Shuttle: Developing an Icon*, 1:208.

84. Jenkins, *Space Shuttle*, 2016, 1:33–41.

85. George Low, "Memorandum Re: Space Shuttle Phase B Statement of Work," August 5, 1969, 1, Box 19, Folder 019884, NASA Historical Reference Collection, NASA Headquarters, Washington, DC.

86. Low, "Memorandum Re: Space Shuttle Phase B Statement of Work," 2.

87. Low, "Memorandum Re: Space Shuttle Phase B Statement of Work," cover note.

88. Low, "Memorandum Re: Space Shuttle Phase B Statement of Work," 1.

3 The Winged Gospel and National Security, 1969–1974

1. Joseph J. Corn, *The Winged Gospel: America's Romance with Aviation, 1900–1950* (New York: Oxford University Press, 1983), 32.

2. Corn, *The Winged Gospel*, 146.

3. Corn, *The Winged Gospel*, 37.

4. For example, see Walter A. McDougall, . . . *the Heavens and the Earth: A Political History of the Space Age* (Baltimore: Johns Hopkins University Press, 1997), 148.

5. John M. Logsdon, *After Apollo? Richard Nixon and the American Space Program* (Palgrave MacMillan, 2015), 32.

6. See, for example, Logsdon, *After Apollo?* 179–180.

7. Logsdon, *After Apollo?* 180 (citing conversation 10, tape 471, March 24, 1971, Richard Nixon Presidential Library and Museum, Yorba Linda, CA).

8. Logsdon, *After Apollo?* 42–46.

9. For example, see Logsdon, *After Apollo?* 258–259.

10. Charles Townes, et al., "Report of the Task Force on Space, January 8, 1969," in *Exploring the Unknown: Selected Documents in the History of the U.S. Civil Space Program*,

ed. John M. Logsdon (Washington, DC: National Aeronautics and Space Administration, 1995), 500–511.

11. Logsdon, *After Apollo?* 241.

12. Seamans, "Letter to Honorable Spiro T. Agnew, Vice President, August 4, 1969," 500.

13. Corn, *The Winged Gospel.*

14. Stephen J. Garber, *Birds of a Feather? How Politics and Culture Affected the Designs of the U.S. Space Shuttle and the Soviet Buran* (master's thesis, Virginia Tech, Blacksburg, 2002).

15. See, generally, Logsdon, *After Apollo?*

16. Howard E McCurdy, *The Space Station Decision: Incremental Politics and Technological Choice*, 1990, 32.

17. Logsdon, *After Apollo?* 69, 77.

18. Michael J. Neufeld, *Von Braun: Dreamer of Space, Engineer of War* (New York: A. A. Knopf, 2007), 438–439.

19. Hans Mark, *The Space Station: A Personal Journey* (Durham, NC: Duke University Press, 1987), 40.

20. Mark, *The Space Station*, 41.

21. Mark, *The Space Station*, 42.

22. Logsdon, *After Apollo?* 163.

23. Thomas O. Paine and Robert C. Seamans, "Memorandum, April 4, 1969, with Attached 'Terms of Reference for Joint DOD/NASA Study of Space Transportation Systems.,'" in *External Relationships*, ed. Dwayne A. Day and John M. Logsdon, vol. 2, Exploring the Unknown: Selected Documents in the History of the U.S. Civil Space Program (Washington, DC: NASA, 1996), 365.

24. Logsdon, *After Apollo?* 70.

25. Seamans, "Letter to Honorable Spiro T. Agnew, Vice President, August 4, 1969," 521.

26. George E. Mueller, "Letter to Kurt H. Debus" (National Aeronautics and Space Administration, September 11, 1969), 2, Folder (George Mueller Papers), NASA Historical Reference Collection, NASA Headquarters, Washington, DC.

27. Logsdon, *After Apollo?* 120.

28. George M. Low, "Memorandum for the Record: 'Space Shuttle Discussions with Secretary Seamans,' January 28, 1970," in *Exploring the Unknown: Selected Documents*

in the History of the U.S. Civil Space Program, vol. 2 (Washington, DC: National Aeronautics and Space Administration, 1996), 367.

29. See, generally, Garber, *Birds of a Feather?*

30. Logsdon, *After Apollo?* 165.

31. Logsdon, 169.

32. R. Dale Reed and Darlene Lister, *Wingless Flight: The Lifting Body Story* (Washington, DC: NASA History Office, Office of Policy and Plans, 1997), 143.

33. Mark, *The Space Station*, 64.

34. Spacecraft launched into polar orbits also require a more powerful rocket than those launched into equatorial orbits, as the Earth's rotation adds velocity to equatorial launches equivalent to several thousand miles per hour.

35. These included Lockheed's experimental, high-speed, high altitude airplane, the A-12 *Oxcart/Cygnus*, as well as its reconnaissance variant, the SR-71 *Blackbird.*

36. T. A. Heppenheimer, *The Space Shuttle Decision, 1965–1972* (Washington, DC: Smithsonian Institution Press, 2002), 270.

37. Logsdon, *After Apollo?* 205–207.

38. Logsdon, *After Apollo?* 186, 188, 194.

39. Charles J. Donlan, "Transmittal of NASA Paper 'Space Shuttle Systems Definition Evolution,' July 11, 1972," in *Exploring the Unknown: Selected Documents in the History of the U.S. Civil Space Program*, vol. IV (Washington, DC: NASA History Division, 1999), 212.

40. Neufeld, *Von Braun: Dreamer of Space, Engineer of War*, 450.

41. Logsdon, *After Apollo?* 194.

42. Thomas O. Paine and Robert C. Seamans Jr., "Agreement between the National Aeronautics and Space Administration and the Department of the Air Force Concerning the Space Transportation System," NMI 1052.130, Attachment A, February 17, 1970," in *External Relationships*, ed. Dwayne A. Day and John M. Logsdon, vol. 2, Exploring the Unknown: Selected Documents in the History of the U.S. Civil Space Program (Washington, DC: NASA, 1996), 368.

43. James C. Fletcher, "The Space Shuttle, November 22, 1971," in *Exploring the Unknown: Selected Documents in the History of the U.S. Civil Space Program*, ed. John M. Logsdon, the NASA History Series (Washington, DC: NASA, 1995), 557.

44. National Air and Space Museum, "NASM Oral History Project, Fletcher," accessed July 23, 2020, https://airandspace.si.edu/research/projects/oral-histories/TRANSCPT/FLETCHER.HTM.

45. Robert C. Truax, "Shuttles—What Price Elegance?" *Astronautics and Aeronautics*, June 1970, 22.

46. Truax, "Shuttles—What Price Elegance?" 22; Roger D. Launius, "The Strange Career of the American Spaceplane: The Long History of Wings and Wheels in Human Space Operations," *Centaurus* 55, no. 4 (November 1, 2013): 421.

47. Robert Truax, "Letter to James C. Fletcher," June 22, 1971, 1, Box 19, Folder 019884, NASA Historical Reference Collection, NASA Headquarters, Washington, DC.

48. Truax, "Letter to James C. Fletcher," 1.

49. Truax, "Letter to James C. Fletcher," 2.

50. Truax, "Letter to James C. Fletcher," 2.

51. George M. Low, "Memorandum Re: Luncheon Conversation with Dave Packard," October 20, 1971, 1, Box 19, Folder 019884, NASA Historical Reference Collection, NASA Headquarters, Washington, DC (handwritten annotations).

52. Low, "Memorandum Re: Luncheon Conversation with Dave Packard," 1–2.

53. George Low, "Memorandum for Dale Myers Re: Space Shuttle Objectives," January 27, 1970, 1, Box 19, Folder 019884, NASA Historical Reference Collection, NASA Headquarters, Washington, DC.

54. Low, "Memorandum Re: Luncheon Conversation with Dave Packard," 1–2.

55. James C. Fletcher, "Memorandum to Associate Administrator for Manned Spaceflight Re: Conversation with Bill Bergen," July 15, 1971, 1, Ms 202, Box 4, Folder 1, James Chipman Fletcher Papers. Special Collections and Archives. University of Utah, J. Willard Marriott. Salt Lake City.

56. Low, "Memorandum Re: Luncheon Conversation with Dave Packard," 1.

57. Mathematica, "Economic Analysis of the Space Shuttle System," January 31, 1972.

58. Mathematica, "Economic Analysis of the Space Shuttle System."

59. Brian O'Leary, "The Space Shuttle: NASA's White Elephant in the Sky," *Bulletin of the Atomic Scientists* 29, no. 2 (February 1973): 36–43.

60. O'Leary, "The Space Shuttle: NASA's White Elephant in the Sky": 37.

61. O'Leary, "The Space Shuttle: NASA's White Elephant in the Sky": 37; See, generally, National Air and Space Museum, "NASM Oral History Project, Fletcher."

62. Heppenheimer, *The Space Shuttle Decision, 1965–1972*, 372–376.

63. Klaus P. Heiss and Oskar Morgenstern, "Memorandum for Dr. James C. Fletcher, Administrator, NASA, 'Factors for a Decision on a New Reusable Space Transportation System,' October 28, 1971," in *Exploring the Unknown: Selected Documents in the*

History of the U.S. Civil Space Program, ed. John M. Logsdon (Washington, DC: NASA, 1995), 552.

64. O'Leary, "The Space Shuttle: NASA's White Elephant in the Sky": 41.

65. Mathematica, "Economic Analysis of the Space Shuttle System," 14.

66. O'Leary, "The Space Shuttle: NASA's White Elephant in the Sky": 37.

67. In Apple TV's fictional, alternative-reality television series *For All Mankind*, the US accelerates the 1960s Cold War Space Race, exploring the Moon and Mars in the 1970s, 1980s, and 1990s to a greater extent than actually occurred. Its most absurd conjectures, though, are economic rather than technological: in the third season, the Soviet Union thrives well into the 1990s, and TAOS proves so profitable that NASA returns funds to the US Treasury. The latter achievement occurs despite the absence of any discernable space commerce; indeed, the first Earth orbital hotel, *Polaris*, is a complete flop. "All In," *For All Mankind* (Apple TV, June 24, 2022).

68. Harold M. Schmenck Jr., "A NASA Aide Sees Industry in Space: Near-Perfect Ball Bearings Could Be Made in Orbit," *New York Times*, November 27, 1968, 44; Gregg Easterbrook, "Beam Me Out of This Death Trap, Scotty," *Washington Monthly*, April 1980; H. Allaway, *The Space Shuttle at Work* (Washington, DC: NASA, 1979), 26; Andrew J. Butrica, *Single Stage to Orbit: Politics, Space Technology, and the Quest for Reusable Rocketry* (Baltimore: Johns Hopkins University Press, 2003), 35; See also Amy Paige Kaminski, *Sharing the Shuttle with America: NASA and Public Engagement after Apollo* (PhD diss., Science and Technology in Society, Blacksburg: Virginia Polytechnic Institute, 2015)

69. Neufeld, *Von Braun: Dreamer of Space, Engineer of War*, 453.

70. George M. Low, "Letter to Donald B. Rice, Assistant Director, Office of Management and Budget, November 22, 1971, with Attached: 'Space Shuttle Configurations,'" in *Accessing Space*, vol. IV, Exploring the Unknown: Selected Documents in the History of the U.S. Civil Space Program (Washington, DC: NASA History Division, 1999), 234.

71. Logsdon, *After Apollo?* 248–249.

72. For example, see Logsdon, *After Apollo?* 241.

73. This section draws from research previously published in modified form. See Matthew H. Hersch, "Using the Shuttle: Operations in Orbit," in *Space Shuttle Legacy: How We Did It and What We Learned*, ed. Roger D. Launius, John Krige, James I. Craig, and Ned Allen, 191–214, Library of Flight Series (Reston, VA: American Institute of Aeronautics and Astronautics, Inc, 2013).

74. John Logsdon, "The Space Shuttle Program: A Policy Failure?" *Science*, no. 232 (May 30, 1986): 1102–1103. See also, for example, George M. Low, "Memorandum to Donald B. Rice, November 22, 1971," in *Accessing Space*, vol. IV, Exploring the

Unknown: Selected Documents in the History of the U.S. Civil Space Program (Washington, DC: NASA History Division, 1999), 231–238; Dennis R. Jenkins, *Space Shuttle: Developing an Icon, 1972–2013*, Vol. 1 (Forest Lake, MN: Specialty Press, 2016): 140–141.

75. John M. Logsdon, "Was the Space Shuttle a Mistake?" *MIT Technology Review*, July 6, 2011, https://www.technologyreview.com/s/424586/was-the-space-shuttle-a-mistake/.

76. In 1968, US intelligence recorded Nixon secretly negotiating with North Vietnam to prolong the war and thereby ruin President Lyndon Johnson's peace initiative and Vice President Hubert Humphrey's election bid. Nixon was convinced that the Democrats would ambush Nixon's reelection campaign with tapes of what Johnson referred to as Nixon's wartime "treason". Obsessed with recovering the recordings, Nixon ordered his staff to undertake a series of burglaries, culminating in the break-in at the Democratic National Committee headquarters in the Watergate Office Building in 1972. At other times, Nixon's staff ignored these illegal orders in the hope that he would forget that he had given them, as when Nixon ordered National Security Advisor Henry Kissinger to burglarize the Brookings Institution six months before the shuttle decision. See, for example, "The Lyndon Johnson Tapes: Richard Nixon's 'Treason,'" *BBC News*, March 22, 2013, sec. Magazine, https://www.bbc.com/news/magazine-21768668.

77. Jenkins, *Space Shuttle: Developing an Icon*, 1: 300–302. See, for example, Arnold R. Weber, "Memorandum for Peter Flanigan, 'Space Shuttle Program,' June 10, 1971, with Attached: 'NASA's Internal Organization for the Space Shuttle Project' and 'NASA's Space Shuttle Program.,'" in *Accessing Space*, vol. IV, Exploring the Unknown: Selected Documents in the History of the U.S. Civil Space Program (Washington, DC: NASA History Division, 1999), 251; James C. Fletcher, George M. Low, and Richard McCurdy, "Memorandum for the Record, 'Selection of Contractor for Space Shuttle Program,' September 18, 1972," in *Accessing Space*, vol. IV, Exploring the Unknown: Selected Documents in the History of the U.S. Civil Space Program (Washington, DC: NASA History Division, 1999), 264.

78. Ben A. Franklin, "Miss Woods's Gift List Links Donors to Corporations," *New York Times*, March 20, 1974, 30; Logsdon, *After Apollo?* 234.

79. James C. Fletcher, "Administrator, NASA, to Caspar W. Weinberger, Deputy Director, Office of Management and Budget, December 29, 1971," in *Accessing Space*, vol. IV, Exploring the Unknown: Selected Documents in the History of the U.S. Civil Space Program (Washington, DC: NASA History Division, 1999), 245.

80. Low, "Memorandum Re: Luncheon Conversation with Dave Packard," 2.

81. Low, "Memorandum Re: Luncheon Conversation with Dave Packard," 1.

82. Jenkins, *Space Shuttle: Developing an Icon*, 1:320 (citing NASA, Appendix to Space Shuttle Facts Sheet, March 15, 1972).

83. James C. Fletcher, "Statement by Dr. Fletcher, NASA Administrator," 1972, https://history.nasa.gov/stsnixon.htm.

84. Jenkins, *Space Shuttle: Developing an Icon*, 1: 314–315.

85. Low, "Memorandum Re: Luncheon Conversation with Dave Packard," 1.

86. Opinion Research Corporation, "Public Knowledge and Attitudes Regarding Space Exploration," September 1972, 44–45, 52–53, Ms 202, Box 10, Folder 6, James Chipman Fletcher Papers. Special Collections and Archives. University of Utah, J. Willard Marriott. Salt Lake City.

87. Richard M. Nixon, "The Statement by President Nixon, 5 January 1972," 1972, http://history.nasa.gov/stsnixon.htm.

88. Maxime A. Faget, "The Space Shuttle," in *Proceedings* (The Space Shuttle: Its Current Status and Future Impact, Warrendale, PA: Society of Automotive Engineers, 1980), 10.

89. Faget, "The Space Shuttle," 7.

90. Faget, "The Space Shuttle," 9.

91. Robert L. Crippen, "Oral History Transcript (Rebecca Wright, Interviewer)," *NASA Johnson Space Center Oral History Project*, 2006, 25.

92. Kenneth S. Thomas and Harold J. McMann, *U.S. Spacesuits* (Chichester, UK: Praxis, 2006), 35–36; Dennis R. Jenkins, *Space Shuttle: Developing an Icon*, Vol. 2 (Forest Lake, MN: Specialty Press, 2016): 373–374.

93. For example, see Dinah L. Moche, *The Star Wars Question and Answer Book about Space* (New York: Random House, 1979).

94. Thomas and McMann, *U.S. Spacesuits*, 37–38.

95. Thomas and McMann, *U.S. Spacesuits*, 37–38; Brian Mork, "Astronaut Job Interview 2000–2009," 2009, http://www.increa.com/NASA2000/; Kevin Cook, *The Burning Blue: The Untold Story of Christa McAuliffe and NASA's Challenger Disaster* (New York: Henry Holt and Company, 2021), 22.

96. Heppenheimer, *The Space Shuttle Decision, 1965–1972*, 165–166.

97. Steven J. Dick, ed., *NASA's First 50 Years: Historical Perspectives*, NASA SP 2010-4704 (Washington, DC: NASA, Office of Communications, History Division, 2010), 340; Michael J. Neufeld, "The 'Von Braun Paradigm' and NASA's Long-Term Planning for Human Spaceflight," in *NASA's First 50 Years: Historical Perspectives*, ed. Steven J. Dick (Washington, DC: NASA, 2010), 325–348.

4 Tickling the Dragon, 1974–1986

1. See, generally, Gabrielle Hecht, *Being Nuclear: Africans and the Global Uranium Trade* (Cambridge, MA: MIT Press, 2012).

2. Robert Serber and Richard Rhodes, *The Los Alamos Primer: The First Lectures on How to Build an Atomic Bomb, Updated with a New Introduction by Richard Rhodes* (University of California Press, 2020), xxxviii.

3. Julian G. West, "The Atomic-Bomb Core That Escaped World War II," https://www.theatlantic.com/technology/archive/2018/04/tickling-the-dragons-tail-plutonium-time-bomb/557006/.

4. Serber and Rhodes, *Los Alamos Primer*, 42.

5. Richard Rhodes, *The Making of the Atomic Bomb* (New York: Simon & Schuster, 1986), 611.

6. Alex Wellerstein, "The Demon Core," *The New Yorker*, May 21, 2016, https://www.newyorker.com/tech/annals-of-technology/demon-core-the-strange-death-of-louis-slotin.

7. See John M. Logsdon, *After Apollo? Richard Nixon and the American Space Program* (Palgrave MacMillan, 2015), 198.

8. John Krige, *NASA in the World: Fifty Years of International Collaboration in Space* (New York: Palgrave Macmillan, 2013), 109 et seq.

9. See Logsdon, *After Apollo?* 198.

10. Krige, *NASA in the World*, 116.

11. Gregg Easterbrook, "Beam Me Out of This Death Trap, Scotty," *Washington Monthly*, April 1980

12. Easterbrook, "Beam Me Out of This Death Trap, Scotty."

13. Tom Campbell, "Note for Dr. Fletcher Re: Quick Look Summary of Spacelab," August 23, 1976, 1, Ms 202, Box 43, Folder 2, James Chipman Fletcher Papers. Special Collections and Archives. University of Utah, J. Willard Marriott. Salt Lake City.

14. Campbell, "Note for Dr. Fletcher," 1.

15. John F. Yardley, "Letter to Director, Public Affairs, NASA, Memorandum, 'Recommended Orbiter Names,' May 26, 1978, with Attached: 'Recommended List of Orbiter Names,'" in *Accessing Space*, vol. IV, Exploring the Unknown: Selected Documents in the History of the U.S. Civil Space Program (Washington, DC: NASA History Division, 1999), 274–275.

16. J. Winston, *The Making of the Trek Conventions: Or, How to Throw a Party for 12,000 of Your Most Intimate Friends* (New York: Doubleday, 1977), 26.

17. Isaac Asimov, "Letter to James C. Fletcher," August 5, 1976, 1, Ms 202, Box 42, Folder 4, James Chipman Fletcher Papers. Special Collections and Archives. University of Utah, J. Willard Marriott. Salt Lake City.

18. Margaret A. Weitekamp, *Space Craze: America's Enduring Fascination with Real and Imagined Spaceflight* (Washington, DC: Smithsonian Books, 2022), 138.

19. Dennis R. Jenkins, *Space Shuttle: Developing an Icon, 1972–2013*, Vol. 1 (Forest Lake, MN: Specialty Press, 2016), 1:422.

20. Recollections of the author.

21. *Moonraker,* Lewis Gilbert (dir.) (Los Angeles: United Artists, 1979); See, generally, Weitekamp, *Space Craze*, 161–162.

22. For example, see Allaway, *Space Shuttle at Work*, 22–23, 31.

23. Allaway, *Space Shuttle at Work*, 4.

24. Allaway, *Space Shuttle at Work*, 51.

25. Government Accounting Office (GAO), "Draft Report to the Congress of the United States, Space Transportation System: Past, Present Future," February 1977, 74, Ms 202, Box 53, Folder 7, James Chipman Fletcher Papers. Special Collections and Archives. University of Utah, J. Willard Marriott. Salt Lake City.

26. For example, see James C. Fletcher, "Letter to Elmer B. Staats," February 7, 1977, 1, Ms 202, Box 53, Folder 5, James Chipman Fletcher Papers. Special Collections and Archives. University of Utah, J. Willard Marriott. Salt Lake City.

27. James C. Fletcher, "Memorandum to Al Lovelace, 'Personal Concern about the Launch Phase of Space Shuttle,' July 7, 1977," in *Accessing Space*, vol. IV, Exploring the Unknown: Selected Documents in the History of the U.S. Civil Space Program (Washington, DC: NASA History Division, 1999), 354.

28. Paul E. Ceruzzi, "Aerospace Needs, Microelectronics, and the Quest for Reliability: 1962–1975," *Proceedings of the IEEE* 105, no. 7 (July 2017): 1456–1465.

29. Chris Petty and Dennis R. Jenkins, *Beyond Blue Skies: The Rocket Plane Programs That Led to the Space Age*, ill. ed. (Lincoln: University of Nebraska Press, 2020), 331.

30. Allaway, *Space Shuttle at Work*, 42.

31. Maxime A. Faget, "The Space Shuttle," in *Proceedings* (The Space Shuttle: Its Current Status and Future Impact, Warrendale, PA: Society of Automotive Engineers, 1980), 14.

32. See, for example, María González Pendás, "Fifty Cents a Foot, 14,500 Buckets: Concrete Numbers and the Illusory Shells of Mexican Economy," *Grey Room* (June 1, 2018): 14–39, https://doi.org/10.1162/grey_a_00240.

33. Hans Mark, *The Space Station: A Personal Journey* (Durham, NC: Duke University Press, 1987).

34. Easterbrook, "Beam Me Out of This Death Trap, Scotty."

35. Allaway, *Space Shuttle at Work*, 1.

36. Office of Management and Budget, "Meeting on the Space Shuttle, November 14, 1979," in *Accessing Space*, vol. IV, Exploring the Unknown: Selected Documents in the History of the U.S. Civil Space Program (Washington, DC: NASA History Division, 1999), 300.

37. See, for example, Katherine P. Van Hooser, "Space Shuttle Main Engine—The Relentless Pursuit of Improvement," 2012, 5. Also see Office of Management and Budget, "Meeting on the Space Shuttle, November 14, 1979," 300.

38. Office of Management and Budget, "Meeting on the Space Shuttle, November 14, 1979," 300.

39. See, for example, Robert Rosenberg, "Why Shuttle Is Needed, Undated but November 1979," in *Accessing Space*, vol. IV, Exploring the Unknown: Selected Documents in the History of the U.S. Civil Space Program (Washington, DC: NASA History Division, 1999), 293.

40. Eric Berger, "A Cold War Mystery: Why Did Jimmy Carter Save the Space Shuttle?" *Ars Technica*, July 14, 2016, https://arstechnica.com/science/2016/07/a-cold-war-mystery-why-did-jimmy-carter-save-the-space-shuttle/; M. D. Damuhn, *Back Down to Earth: The Development of Space Policy for NASA during the Jimmy Carter Administration* (University of Wales Swansea, 2000).

41. Shortly after the liftoff of the Saturn V launch vehicle that carried Skylab into orbit, the station's protective shroud experienced unexpected aerodynamic loads that tore off one its two large solar panels, jammed the other, and ripped off the station's micrometeoroid and solar shield. Repairs by Apollo crews restored the station's functionality, and Skylab ultimately exceeded expectations. See William David Compton and Charles D. Benson, *Living and Working in Space: A History of Skylab* (Washington, DC: Scientific and Technical Information Branch, NASA, 1983); David J. Shayler, *Skylab: America's Space Station* (Chichester, UK: Praxis, 2001).

42. Compton and Benson, *Living and Working in Space*.

43. National Aeronautics and Space Administration (NASA), "STS Quarterly Review, Space Shuttle Program," October 27, 1976, 96SSV51579 et seq., Ms 202, Box 43, Folder 3, James Chipman Fletcher Papers. Special Collections and Archives. University of Utah, J. Willard Marriott. Salt Lake City.

44. National Aeronautics and Space Administration (NASA), Lyndon B. Johnson Space Center, "Major Safety Concerns: Space Shuttle Program, JSC 09990C, November 8, 1976, Preface and Pp. 1-1-2-3, 5–1–A-6.," in *Accessing Space*, vol. IV, Exploring the Unknown: Selected Documents in the History of the U.S. Civil Space Program (Washington, DC: NASA History Division, 1999), 323.

45. See, for example, Matthew H. Hersch, "'Capsules Are Swallowed': The Mythology of the Pilot in American Spaceflight," in *Spacefarers: Images of Astronauts and*

Cosmonauts in the Heroic Era of Spaceflight, ed. Michael J. Neufeld (Washington, DC: Smithsonian Institution, 2013); Matthew H. Hersch, *Inventing the American Astronaut*, Palgrave Studies in the History of Science and Technology (New York: Palgrave Macmillan, 2012).

46. NASA's initial nomenclature assigned mission numbers to shuttle flights in chronological order of launch, regardless of their mission or payload. As NASA began to juggle mission schedules, though, planning for future flights became unwieldy, and by the end of 1984, the agency replaced the simple numerical naming scheme with a number-and-letter code matched to the mission's payload rather than its position in the launch schedule. The code proved so confusing, though, that almost no one outside NASA could keep track of flights. NASA returned to the old system after the loss of *Challenger* in 1986. Weitekamp, *Space Craze*, 184.

47. Robert L. Crippen, "Oral History Transcript (Rebecca Wright, Interviewer)," *NASA Johnson Space Center Oral History Project*, 2006, 26.

48. See, for example, John F. Yardley, "Study of TPS Inspection and Repair On-Orbit, June 14, 1979," in *Accessing Space*, vol. IV, Exploring the Unknown: Selected Documents in the History of the U.S. Civil Space Program (Washington, DC: NASA History Division, 1999), 282.

49. Crippen, "Oral History Transcript (Rebecca Wright, Interviewer)," 27.

50. Rowland White, *Into the Black: The Extraordinary Untold Story of the First Flight of the Space Shuttle Columbia and the Men Who Flew Her* (New York: Touchstone, 2016).

51. Joseph P. Allen, "Oral History Transcript #3 (Jennifer Ross-Nazzal, Interviewer)," *NASA Johnson Space Center Oral History Project*, 2004, 11.

52. John M. Logsdon, *Ronald Reagan and the Space Frontier*, Palgrave Studies in the History of Science and Technology (Cham, Switzerland: Springer International Publishing AG, 2018), 71.

53. Logsdon, *Ronald Reagan and the Space Frontier*, 171.

54. Ronald W. Reagan, "Inaugural Address," Ronald Reagan Presidential Library and Museum, 1981, https://www.reaganlibrary.gov/archives/speech/inaugural-address-1981.

55. Logsdon, *Ronald Reagan and the Space Frontier*, 7.

56. Logsdon, *Ronald Reagan and the Space Frontier*, 35.

57. Logsdon, *Ronald Reagan and the Space Frontier*, 25.

58. Reagan, "Inaugural Address," 20–22.

59. Logsdon, *Ronald Reagan and the Space Frontier*, 192.

60. Logsdon, *Ronald Reagan and the Space Frontier*, 171–172.

61. L. Cywanowicz, "NASA Expands Payload Specialist Opportunities," *NASA News (Marshall Space Flight Center)*, 1982, 1.

62. James M. Beggs, "Letter to Don Fuqua," October 7, 1982, Folder 008960, NASA Historical Reference Collection, NASA Headquarters, Washington, DC.

63. Logsdon, *Ronald Reagan and the Space Frontier*, 74–75.

64. Logsdon, *Ronald Reagan and the Space Frontier*, 177.

65. See, for example, Diane Vaughan, *The Challenger Launch Decision: Risky Technology, Culture, and Deviance at NASA* (Chicago: University of Chicago Press, 1996); John Logsdon, "The Space Shuttle Program: A Policy Failure?" *Science*, no. 232 (May 30, 1986).

66. For example, see Brian O'Leary, "The Space Shuttle: NASA's White Elephant in the Sky," *Bulletin of the Atomic Scientists* 29, no. 2 (February 1973): 42. See, also, e.g., Otto K. Goetz, "Edited Oral History Transcript (Jennifer Rozz-Nazzal, Interviewer)," *NASA STS Recordation Oral History Project*, 2000, 12. See, also, e.g., Otto K. Goetz, "Edited Oral History Transcript (Jennifer Rozz-Nazzal, Interviewer)," *NASA STS Recordation Oral History Project*, 2000, 12.

67. Mike Mullane, *Riding Rockets: The Outrageous Tales of a Space Shuttle Astronaut* (New York: Scribner, 2006), 206.

68. Donald K. Slayton and Michael Cassutt, *Deke! U.S. Manned Space: From Mercury to the Shuttle* (New York: St. Martin's Press, 1994), 316.

69. Allen, "Oral History Transcript #3 (Jennifer Ross-Nazzal, Interviewer)," 30–31.

70. Asif A. Siddiqi, *Challenge to Apollo: The Soviet Union and the Space Race, 1945–1974* (Washington, DC: NASA, NASA History Division, Office of Policy and Plans, 2000), 781.

71. Siddiqi, *Challenge to Apollo*, 703.

72. Crippen, "Oral History Transcript (Rebecca Wright, Interviewer)," 43.

73. C. G. Fullerton, *Oral History Transcript* (Rebecca Wright, Interviewer) (NASA Dryden Flight Research Center: NASA, 2002), 37.

74. Crippen, "Oral History Transcript (Rebecca Wright, Interviewer)," 31.

75. Crippen, 43. See also *Vanity Fair*, "Should Airplanes Be Flying Themselves?" September 17, 2014, https://www.vanityfair.com/news/business/2014/10/air-france-flight-447-crash.

76. Crippen, "Oral History Transcript (Rebecca Wright, Interviewer)," 74.

77. Allen, "Oral History Transcript #3 (Jennifer Ross-Nazzal, Interviewer)," 8–9.

78. See, for example, Matthew H. Hersch, "Using the Shuttle: Operations in Orbit," in *Space Shuttle Legacy: How We Did It and What We Learned*, ed. Roger D. Launius,

John Krige, James I. Craig, and Ned Allen, 191–214, Library of Flight Series (Reston, VA: American Institute of Aeronautics and Astronautics, Inc, 2013).

79. Michael Cassutt, "Secret Space Shuttles," *Air & Space*, 2009; see also, generally, James E. David, *Spies and Shuttles: NASA's Secret Relationships with the DOD and CIA* (Gainesville: University Press of Florida, 2015).

80. Cassutt, "Secret Space Shuttles," 45–46; see also, generally, David, *Spies and Shuttles*.

81. Allen, "Oral History Transcript #3 (Jennifer Ross-Nazzal, Interviewer)," 32; See also Michael Cassutt, *The Astronaut Maker: How One Mysterious Engineer Ran Human Spaceflight for a Generation* (Chicago: Review Press, 2018), 244.

82. Donald H. Peterson, "Oral History Transcript (Jennifer Ross-Nazzal, Interviewer)," *NASA Johnson Space Center Oral History Project*, 2002, 63; Chester Lee, "Letter to Manager, Space Shuttle Payload Integration and Development Program Office, Johnson Space Center, and Manager, STS Projects Office, Kennedy Space Center, 'Guidelines for Development of the Flight Assignment Baseline,' November 20, 1978," in *Accessing Space*, vol. IV, Exploring the Unknown: Selected Documents in the History of the U.S. Civil Space Program (Washington, DC: NASA History Division, 1999), 346.

83. Robert Schulman, "NASA Art Program: Space Shuttle Art Collection" (NASA, n.d.), 2, Box 1 Folder ("The Artist and the Shuttle: Organizer"), Smithsonian Institution Traveling Exhibition Service, Exhibition Records, c. 1979–1995, Smithsonian Institution Archives.

84. "News Release: Space Shuttle Art Exhibition to Go on View at the National Air and Space Museum" (Smithsonian Institution, November 12, 1982), Box 1 Folder ("The Artist and the Shuttle: Press"), Smithsonian Institution Traveling Exhibition Service (SITES), Exhibition Records, c. 1979–1995, Smithsonian Institution Archives; SITES, "The Artist and the Space Shuttle: Itinerary" (Smithsonian Institution, November 12, 1982), Box 1 Folder ("The Artist and the Shuttle: Press"), Smithsonian Institution Traveling Exhibition Service, Exhibition Records, c. 1979–1995, Smithsonian Institution Archives; National Aeronautics and Space Administration, Public Affairs Division, "Our First Quarter Century of Achievement . . . Just the Beginning," Press Release (Washington, DC, October 1983), L-5; "The Artist and the Space Shuttle," *Discover*, December 1982.

85. Robert Rauschenberg, *Hot Shot*, 1983, https://artmuseum.williams.edu/collection/featured-acquisitions/robert-rauschenberg/; "Stoned Moon (1969–70)," Robert Rauschenberg Foundation, July 11, 2013, https://www.rauschenbergfoundation.org/art/series/stoned-moon-edition; "Stoned Moon," Robert Rauschenberg Foundation, December 22, 2014, https://www.rauschenbergfoundation.org/art/art-in-context/stoned-moon.

86. "Statement on Hot Shot (1983)," Robert Rauschenberg Foundation, September 4, 2014, https://www.rauschenbergfoundation.org/art/archive/a9.

87. Bert Ulrich, "NASA and the Arts," NASA.gov (Brian Dunbar, July 3, 2013), https://www.nasa.gov/50th/50th_magazine/arts.html.

88. *Man in Space.*

89. Compton and Benson, *Living and Working in Space*; Shayler, *Skylab: America's Space Station.*

90. Barton C. Hacker and James M. Grimwood, *On the Shoulders of Titans: A History of Project Gemini* (Washington, DC: NASA, 1978).

91. Allaway, *Space Shuttle at Work*, 68.

92. Overworked, underfed, and underslept, the three astronauts of the third Skylab crew likely disabled their radio briefly on December 28, 1973, to avoid receiving further instructions from ground controllers. Concerned, NASA relented, allowing the astronauts a day off to catch up on past work and engage in recreation. The agency later grounded the men permanently; they never flew in space again. Compton and Benson, *Living and Working in Space*, 326–330.

93. David J. Shayler, *NASA's Scientist-Astronauts* (New York: Springer, 2007), 336.

94. For example, see Hersch, *Inventing the American Astronaut*, 150–51.

95. Philip M. Boffey, "NASA to Seek Observers to Fly on Shuttles," *New York Times*, December 16, 1983, 11.

96. Cassutt, *The Astronaut Maker*, 279.

97. See Kaminski, *Sharing the Shuttle with America,* chap. 4.

98. Kevin Cook, *The Burning Blue: The Untold Story of Christa McAuliffe and NASA's Challenger Disaster* (New York: Henry Holt and Company, 2021), 13.

99. Philip M. Boffey, "First Shuttle Ride by Private Citizen to Go to Teacher: Hopes of Others Dashed," *New York Times*, August 28, 1984, A1.

100. Boffey, "First Shuttle Ride by Private Citizen."

101. K. J. Bobko, "Oral History Transcript (Summer Chick Bergen, Interviewer)," *NASA Johnson Space Center Oral History Project*, 2002, 33.

102. See Cassutt, *The Astronaut Maker*, 277.

103. Allen, "Oral History Transcript #4 (Jennifer Ross-Nazzal, Interviewer)," 12.

104. Cassutt, *The Astronaut Maker*, 272.

105. Henry W. Hartsfield, Jr., "Oral History Transcript #2 (Carol Butler, Interviewer)," *NASA Johnson Space Center Oral History Project*, 2001, 26.

106. Hartsfield, "Oral History Transcript #2 (Carol Butler, Interviewer)," 26.

107. Alex Roland, "Triumph or Turkey?" *Discover*, 1985.

108. Robert T. Hohler, *I Touch the Future: The Story of Christa McAuliffe* (New York: Random House, 1986), 150.

109. R. Dale Reed and Darlene Lister, *Wingless Flight: The Lifting Body Story* (Washington, DC: NASA History Office, Office of Policy and Plans, 1997), xxv.

110. John Noble Wilford, "A Teacher Trains for Outer Space," *New York Times*, January 5, 1986, SM16; Cook, *The Burning Blue*, chap. 2.

111. Cook, *The Burning Blue*, 25.

112. Hohler, *I Touch the Future*, 148–149.

113. Cook, *The Burning Blue*, 42.

114. Bettyann H. Kevles, *Almost Heaven: The Story of Women in Space* (New York: Basic Books, 2003), 105–106.

115. Hartsfield, "Oral History Transcript (Carol Butler, Interviewer)," 40–41.

116. Mullane, *Riding Rockets*, 206.

117. Andrew J. Butrica, *Single Stage to Orbit: Politics, Space Technology, and the Quest for Reusable Rocketry* (Baltimore: Johns Hopkins University Press, 2003), 34.

5 A History Rooted in Accident, 1986–2011

1. Thomas O'Toole, "Glamor Is Dying in the Camelot of the Astronauts," *The Washington Post*, February 5, 1986, Final edition.

2. Donald H. Peterson, "Oral History Transcript (Jennifer Ross-Nazzal, Interviewer)," *NASA Johnson Space Center Oral History Project*, 2002, 55.

3. Diane Vaughan, *The Challenger Launch Decision: Risky Technology, Culture, and Deviance at NASA* (Chicago: University of Chicago Press, 1996), 153, et seq.

4. See, generally, Philip Chien, *Columbia, Final Voyage: The Last Flight of NASA's First Space Shuttle* (New York: Copernicus Books, 2006).

5. See, generally, Vaughan, *The Challenger Launch Decision*; Allan J. McDonald, *Truth, Lies, and O-Rings: Inside the Space Shuttle Challenger Disaster* (Gainesville: University Press of Florida, 2009); Kevin Cook, *The Burning Blue: The Untold Story of Christa McAuliffe and NASA's Challenger Disaster* (New York: Henry Holt and Company, 2021).

6. Rogers Commission, *Report of the Presidential Commission on the Space Shuttle Challenger Accident*, vol. 1 (Washington, DC: Government Printing Office, 1986).

7. Vaughan, *The Challenger Launch Decision*, 10.

8. Peterson, "Oral History Transcript (Jennifer Ross-Nazzal, Interviewer)," 55; Sally K. Ride, "Two Small Notebooks Containing Ride's Notes from the Rogers Commission,"

1986, 15, NASM-NASM.2014.0025-bx013-fd008_001, Sally K. Ride Papers, National Air and Space Museum Archives, Washington, DC.

9. See, generally, Cook, *The Burning Blue*.

10. Rogers Commission, *Report of the Presidential Commission on the Space Shuttle Challenger Accident*.

11. Stanley Karnow, *Vietnam: A History* (New York: Viking Press, 1983), 587–588.

12. Cook, *The Burning Blue*, 213.

13. See, for example, *Challenger: The Final Flight*, Daniel Junge and Steven Leckart (dirs..) (Santa Monica, CA, Melbourne, FL, Culver City, CA: Bad Robot, Zipper Bros Films, Sutter Road Picture Company, 2020).

14. Vaughan, *The Challenger Launch Decision*, 7.

15. Donald K. Slayton and Michael Cassutt, *Deke! U.S. Manned Space: From Mercury to the Shuttle* (New York: St. Martin's Press, 1994), 318.

16. John W. Young, "One Part of the 51-L Accident—Space Shuttle Program Flight Safety, March 4, 1986, with Attached: 'Examples of Uncertain Operational and Engineering Conditions or Events Which We "Routinely" Accept Now in the Space Shuttle Program,'" in *Exploring the Unknown: Selected Documents in the History of the U.S. Civil Space Program*, vol. IV (Washington, DC: NASA History Division, 1999), 379.

17. Young, "One Part of the 51-L Accident ," 379.

18. Rogers Commission, *Report of the Presidential Commission on the Space Shuttle Challenger Accident*.

19. US House Committee on Science and Technology, "Investigation of the Challenger Accident" (Washington, DC: Government Printing Office, October 29, 1986), https://www.govinfo.gov/content/pkg/GPO-CRPT-99hrpt1016/pdf/GPO-CRPT-99hrpt1016.pdf.

20. US House Committee on Science and Technology, "Investigation of the Challenger Accident," 4.

21. Richard P. Feynman, *"What Do You Care What Other People Think?": Further Adventures of a Curious Character* (New York: W. W. Norton, 2018).

22. McDonald, *Truth, Lies, and O-Rings*; Phillip K. Tompkins and Emily V. Tompkins, *Apollo, Challenger, Columbia: The Decline of the Space Program: A Study in Organizational Communication* (Los Angeles: Roxbury Publishing, 2005); Edward R. Tufte, *The Cognitive Style of PowerPoint* (Cheshire, UK: Graphics Press, 2003); Howard E. McCurdy, *Inside NASA: High Technology and Organizational Change in the U.S. Space Program* (Baltimore: Johns Hopkins University Press, 1993).

23. Charles B. Fleddermann, *Engineering Ethics*, 4th ed. (Upper Saddle River, NJ: Pearson, 2011).

24. Vaughan, *The Challenger Launch Decision*, xv.

25. Hannah Arendt, *Eichmann in Jerusalem: A Report on the Banality of Evil* (New York: Viking Press, 1963); Vaughan, *The Challenger Launch Decision*, 17.

26. Vaughan, *The Challenger Launch Decision*, xiv, 17.

27. Similar volumes followed the loss of *Columbia* during reentry in 2003. For example, see Michael Cabbage, *Comm Check . . . : The Final Flight of Shuttle Columbia* (New York: Free Press, 2004); Chien, *Columbia, Final Voyage.*

28. Vaughan, *The Challenger Launch Decision*, xiv.

29. Rogers Commission, *Report of the Presidential Commission on the Space Shuttle Challenger Accident*, 1:120.

30. Vaughan, *The Challenger Launch Decision*, 62.

31. Vaughan, *The Challenger Launch Decision*, 17.

32. Vaughan, *The Challenger Launch Decision*, 34, 415; Charles Perrow, *Normal Accidents: Living with High-Risk Technologies* (New York: Basic Books, 1984).

33. Charles Perrow, *Normal Accidents: Living with High-Risk Technologies: With a New Afterword and a Postscript on the Y2K Problem* (Princeton, NJ: Princeton University Press, 1999), 379–380.

34. Perrow, *Normal Accidents: With a New Afterword and a Postscript on the Y2K Problem*, 98.

35. Andrew Hopkins, "Was Three Mile Island a 'Normal Accident'?" *Journal of Contingencies and Crisis Management* 9, no. 2 (2002): 65–72, https://doi.org/10.1111/1468-5973.00155.

36. Arendt's application of the concept to the case of the Holocaust perpetrator Adolf Eichmann (whom she argued was a mindless bureaucrat consumed by National Socialist culture), has been similarly criticized for neglecting evidence of Eichmann's virulent anti-Semitism. Arendt also worked mostly from transcripts of the trial rather than direct observation of participants. For example, see Bettina Stangneth, *Eichmann before Jerusalem: The Unexamined Life of a Mass Murderer* (New York: Knopf Doubleday Publishing Group, 2014). Also see Perrow, *Normal Accidents: With a New Afterword and a Postscript on the Y2K Problem*, 98.

37. Perrow, *Normal Accidents: With a New Afterword and a Postscript on the Y2K Problem*, 379–380.

38. Harry M. Collins and Trevor J. Pinch, *The Golem at Large: What You Should Know about Technology* (New York: Cambridge University Press, 1998), 38.

39. Collins and Pinch, *The Golem at Large*, 54.

40. Collins and Pinch, *The Golem at Large*, 32.

41. Collins and Pinch, *The Golem at Large*, 55.

42. Collins and Pinch, *The Golem at Large*, 36.

43. US House Committee on Science and Technology, "Investigation of the Challenger Accident."

44. Ride, "Two Small Notebooks Containing Ride's Notes from the Rogers Commission," 15.

45. Vaughan, *The Challenger Launch Decision*, 97.

46. Vaughan, *The Challenger Launch Decision*, 97.

47. George M. Low, "Memorandum Re: Meeting with Sam Phillips," February 28, 1972, 1, Box 19, Folder 019884, NASA Historical Reference Collection, NASA Headquarters, Washington, DC.

48. Vaughan, *The Challenger Launch Decision*, Appendix A.

49. Vaughan, *The Challenger Launch Decision*, Appendix A.

50. Vaughan, *The Challenger Launch Decision*, 100.

51. Dennis R. Jenkins, *Space Shuttle: Developing an Icon, 1972–2013,* Vol. 2 (Forest Lake, MN: Specialty Press, 2016), 417–424.

52. Michael J. Neufeld, *Von Braun: Dreamer of Space, Engineer of War* (New York: A. A. Knopf, 2007), 450.

53. See, generally, Dennis R. Jenkins, *Space Shuttle: Developing an Icon, 1972–2013,* Vol. 1 (Forest Lake, MN: Specialty Press, 2016), 2016

54. Vaughan, *The Challenger Launch Decision*, 22.

55. Hopkins, "Was Three Mile Island a 'Normal Accident'?" 72.

56. Vaughan, *The Challenger Launch Decision*, 110.

57. Vaughan, *The Challenger Launch Decision*, 133.

58. Cook, *The Burning Blue*, 202.

59. Vaughan, *The Challenger Launch Decision*, 189.

60. Vaughan, *The Challenger Launch Decision*, 227, 242.

61. For example, see James C. Fletcher, "Letter to Elmer B. Staats," February 7, 1977, 1, Ms 202, Box 53, Folder 5, James Chipman Fletcher Papers. Special Collections and Archives. University of Utah, J. Willard Marriott. Salt Lake City, 1; Neufeld, *Von Braun:*

Dreamer of Space, Engineer of War, 450; Charles J. Donlan, "Transmittal of NASA Paper 'Space Shuttle Systems Definition Evolution,' July 11, 1972," in *Exploring the Unknown: Selected Documents in the History of the U.S. Civil Space Program*, vol. IV (Washington, DC: NASA History Division, 1999), 212.

62. For example, see Klaus P. Heiss and Oskar Morgenstern, "Memorandum for Dr. James C. Fletcher, Administrator, NASA, 'Factors for a Decision on a New Reusable Space Transportation System,' October 28, 1971," in *Exploring the Unknown: Selected Documents in the History of the U.S. Civil Space Program*, ed. John M. Logsdon (Washington, DC: NASA, 1995), 552.

63. Cook, *The Burning Blue*, 111.

64. Government Accounting Office (GAO), "Draft Report to the Congress of the United States, Space Transportation System: Past, Present Future," 74.

65. Cook, *The Burning Blue*, 110.

66. Vaughan, *The Challenger Launch Decision*, 28.

67. Vaughan, *The Challenger Launch Decision*, 28.

68. See, generally, Feynman, *"What Do You Care What Other People Think?"*; Ride, "Two Small Notebooks Containing Ride's Notes from the Rogers Commission," 151; Margaret Lazarus Dean, "It's the 36th Anniversary of the Space Shuttle Challenger Disaster," *Popular Mechanics*, January 28, 2022, https://www.popularmechanics.com/space/a18616/an-oral-history-of-the-space-shuttle-challenger-disaster/.

69. For example, see Vaughan, *The Challenger Launch Decision*, 274.

70. Cook, *The Burning Blue*, 70; Arnold Barnett, "Aviation Safety: A Whole New World?" *Transportation Science* 54, no. 1 (January 1, 2020): 89; "Commercial Air Travel Passenger Death Rate Continues to Decline: MIT Study," *Insurance Journal*, January 27, 2020, https://www.insurancejournal.com/news/national/2020/01/27/556562.htm.

71. Richard P. Feynman, "Appendix F: Personal Observations on the Reliability of the Shuttle," in *Report of the Presidential Commission on the Space Shuttle Challenger Accident* (Washington, DC: Government Printing Office, 1986).

72. Collins and Pinch, *The Golem at Large*, 55–56.

73. H. Guyford Stever, "Letter from Panel on Redesign of Space Shuttle Solid Rocket Booster, Committee on NASA Scientific and Technological Program Reviews, National Research Council, to James C. Fletcher, Administrator, NASA, Seventh Interim Report, September 9, 1986," in *Exploring the Unknown: Selected Documents in the History of the U.S. Civil Space Program*, vol. IV (Washington, DC: NASA History Division, 1999), 385–393.

74. Revised abort procedures, for example, would still require the crew of a Challenger-like accident to move around within their falling, tumbling crew compartment,

activate various systems, remove hatches, bail out, and pull rip-cords on parachutes—a near impossibility. David Pogue, "Escape Refresher 41020, Space Shuttle Crew Escape Briefing Handout, Undated," n.d., NASM-NASM.2006.0013-bx006-fd012_010, David M. Brown Papers, National Air and Space Museum Archives, Washington, DC.

75. Jenkins, *Space Shuttle: Developing an Icon*, 2: 422

76. For example, see Amy Paige Kaminski, *Sharing the Shuttle with America: NASA and Public Engagement after Apollo* (PhD diss., Science and Technology in Society, Blacksburg: Virginia Polytechnic Institute, 2015), chap. 5.

77. Donald H. Heaton, "Memorandum for the Files," August 23, 1960, Folder 008211, NASA Historical Reference Collection, NASA Headquarters, Washington, DC"; John M. Logsdon, *Ronald Reagan and the Space Frontier*, Palgrave Studies in the History of Science and Technology (Cham, Switzerland: Springer International Publishing AG, 2018), 217.

78. Dennis Eskow, "Space Vacation 1995," *Popular Mechanics*, 1985; "Next: Earth-Orbiting Vacations!" *Christian Science Monitor*. October 19, 1968.

79. Howard E. McCurdy, *Space and the American Imagination* (Washington, DC: Smithsonian Institution Press, 1997), 84.

80. This section draws from research published earlier. See Matthew H. Hersch, "Using the Shuttle: Operations in Orbit," in *Space Shuttle Legacy: How We Did It and What We Learned*, ed. Roger D. Launius, John Krige, James I. Craig, and Ned Allen, 191–214, Library of Flight Series (Reston, VA: American Institute of Aeronautics and Astronautics, Inc, 2013).

81. Smith, *The Space Telescope*, 243.

82. Smith, *The Space Telescope*, 75.

83. Editorial Board, "Bean Counter: O'Keefe Offers Case Study of NASA Don't," *Houston Chronicle*, December 19, 2004, https://www.chron.com/opinion/editorials/article/Bean-counter-O-Keefe-offers-case-study-of-NASA-1970055.php.

84. The author thanks David DeVorkin for this insight.

85. Columbia Accident Investigation Board (CAIB), *The Columbia Accident Investigation Board Report*, vol. 1 (Washington, DC: Government Printing Office, 2003), 127.

86. Columbia Accident Investigation Board (CAIB), *The Columbia Accident Investigation Board Report*, 127.

87. Columbia Accident Investigation Board (CAIB), *The Columbia Accident Investigation Board Report*,,55.

88. Columbia Accident Investigation Board (CAIB), *The Columbia Accident Investigation Board Report*, 60.

89. Columbia Accident Investigation Board (CAIB), *The Columbia Accident Investigation Board Report*, 127.

90. Columbia Accident Investigation Board (CAIB), *The Columbia Accident Investigation Board Report*, 5.

91. Columbia Accident Investigation Board (CAIB), *The Columbia Accident Investigation Board Report*, 64–68.

92. Columbia Accident Investigation Board (CAIB), *The Columbia Accident Investigation Board Report*.

93. See, generally, Vaughan, *The Challenger Launch Decision*.

94. Columbia Accident Investigation Board (CAIB), *The Columbia Accident Investigation Board Report*, 126.

95. Erica Robles-Anderson and Patrik Svensson, "'One Damn Slide after Another': PowerPoint at Every Occasion for Speech," *Computational Culture*, no. 5 (January 15, 2016), http://computationalculture.net/one-damn-slide-after-another-powerpoint-at-every-occasion-for-speech/.

96. Columbia Accident Investigation Board (CAIB), "Engineering by Viewgraph," in *The Columbia Accident Investigation Board Report* (Washington, DC: US Government Printing Office, 2003), 191.

97. Tufte, *Cognitive Style of PowerPoint*.

98. Columbia Accident Investigation Board (CAIB), *The Columbia Accident Investigation Board Report*, 25.

99. Tim Wilson, "External Tank Tiger Team Report," 2005, 80.

100. For example, see Arjen Boin and Paul Schulman, "Assessing NASA's Safety Culture: The Limits and Possibilities of High-Reliability Theory," *Public Administration Review* 68, no. 6 (2008): 1050–1062.

101. Columbia Accident Investigation Board (CAIB), *The Columbia Accident Investigation Board Report*, 127.

102. Mike Mullane, *Riding Rockets: The Outrageous Tales of a Space Shuttle Astronaut* (New York: Scribner, 2006), 282.

103. William Harwood, "STS-119 Shuttle Report: Legendary Commander Tells Story of Shuttle's Close Call," *Spaceflight Now*, March 27, 2009, https://spaceflightnow.com/shuttle/sts119/090327sts27/; Mullane, *Riding Rockets*, 282.

104. Perrow, *Normal Accidents: Living with High-Risk Technologies*.

105. See, for example, Lee Hutchinson, "The Audacious Rescue Plan That Might Have Saved Space Shuttle Columbia," *Ars Technica*, February 1, 2016, https://arstechnica

.com/science/2016/02/the-audacious-rescue-plan-that-might-have-saved-space-shuttle-columbia/.

106. Columbia Accident Investigation Board (CAIB), "STS-107 In-Flight Options Assessment," in *The Columbia Accident Investigation Board Report*, vol. 2 (Washington, DC: Government Printing Office, 2003).

107. Warren E. Leary, "NASA Leader Cites Finances and Submits His Resignation," *New York Times*, December 14, 2004, 20; Editorial Board, "Bean Counter."

108. Wayne Hale, "How We Nearly Lost Discovery," *Wayne Hale's Blog*, April 18, 2012, https://waynehale.wordpress.com/2012/04/18/how-we-nearly-lost-discovery/.

109. For example, see Christine Emba, "Facebook Isn't Too Big to Fail," *Washington Post*, April 10, 2018, https://www.washingtonpost.com/opinions/facebook-isnt-too-big-to-fail/2018/04/10/dd9e9c18-3d02-11e8-a7d1-e4efec6389f0_story.html.

110. Quora, "Why Did NASA End the Space Shuttle Program?" *Forbes*, accessed October 21, 2022, https://www.forbes.com/sites/quora/2017/02/02/why-did-nasa-end-the-space-shuttle-program/.

111. National Aeronautics and Space Administration (NASA), "President Bush Delivers Remarks on U.S. Space Policy," *NASA Facts*, January 14, 2004, 4.

6 The Quest for Alternatives, 1972–2011

1. *Starflight: The Plane That Couldn't Land* (ABC, February 27, 1983).

2. *SpaceCamp*, Harry Winer (dir.) (Twentieth Century Fox, 1986); See, generally, Margaret A. Weitekamp, *Space Craze: America's Enduring Fascination with Real and Imagined Spaceflight* (Washington, DC: Smithsonian Books, 2022), 174.

3. See Weitekamp, *Space Craze*, 166–167.

4. For example, see Paul Attanasio, "Movies," *Washington Post*, June 6, 1986, sec. D.

5. *Armageddon*, Michael Bay (dir.). (Los Angeles: Touchstone Pictures, 1998).

6. *Space Cowboys* Clint Eastwood (dir.) (Los Angeles: Warner Bros., 2000).

7. *Mission to Mars*, Brian de Palma (dir.). (Los Angeles: Buena Vista Distribution, 2000.

8. *The Core*, Jon Amiel (dir.) (Los Angeles: Paramount Pictures, 2003).

9. *Gravity*, Alfonso Cuarón (dir.) (Los Angeles: Warner Bros., 2013).

10. Roger D. Launius and Howard E. McCurdy, eds., "Something Borrowed, Something Blue: Repurposing NASA's Spacecraft," in *NASA Spaceflight: A History of Innovation* (New York: Palgrave Macmillan, 2018), 215–235.

11. See, generally, Matthew H. Hersch, "Using the Shuttle: Operations in Orbit," in *Space Shuttle Legacy: How We Did It and What We Learned*, ed. Roger D. Launius, John

Krige, James I. Craig, and Ned Allen, 191–214, Library of Flight Series (Reston, VA: American Institute of Aeronautics and Astronautics, Inc, 2013).

12. *For All Mankind,* mentioned earlier in this book, is one cinematic effort to imagine what such a future might have looked like, engendered by a competitive geopolitical environment and a determination to continue the use of Apollo hardware simultaneously with shuttle development. *For All Mankind,* Al Reinert (dir.). (Sony Pictures Television, Tall Ship Productions, 2019).

13. John M. Logsdon, *After Apollo? Richard Nixon and the American Space Program* (Palgrave MacMillan, 2015), 203–205.

14. See, generally, Edward C. Ezell and Linda N. Ezell, *The Partnership: A History of the Apollo-Soyuz Test Project* (Washington, DC: NASA, 1978).

15. See, for example, Andrew L. Jenks, *Collaboration in Space and the Search for Peace on Earth,* Anthem Series on Russian, East European and Eurasian Studies (London: Anthem Press, 2022), 76 et seq.

16. William David Compton and Charles D. Benson, *Living and Working in Space: A History of Skylab* (Washington, DC: Scientific and Technical Information Branch, NASA, 1983), 116–118; David J. Shayler, *Skylab: America's Space Station* (Chichester, UK: Praxis, 2001); David J. Shayler, *Apollo: The Lost and Forgotten Missions* (Chichester, UK: Springer, 2002).

17. For example, see "Skylab B," 2019, http://www.astronautix.com/s/skylabb.html.

18. David S. F. Portree, *Humans to Mars: Fifty Years of Mission Planning, 1950–2000* (Washington, DC: NASA History Division, Office of Policy and Plans, 2001), 61.

19. J. Donnelly, "Letter to James Fletcher," George M. Low Papers, NASA HQ, Folder 004157, 1973.

20. Asif A. Siddiqi, *Challenge to Apollo: The Soviet Union and the Space Race, 1945–1974* (Washington, DC: NASA, NASA History Division, Office of Policy and Plans, 2000), 836.

21. Siddiqi, *Challenge to Apollo,* 601, et seq.

22. See, for example, Boris Chertok, *The Moon Race,* vol. IV, Rockets and People, ed. Asif Siddiqi (Washington, DC: US Government Printing Office, 2011).

23. See, for example, Stephen J. Garber, *Birds of a Feather? How Politics and Culture Affected the Designs of the U.S. Space Shuttle and the Soviet Buran* (master's thesis, Virginia Tech, Blacksburg, 2002), 17; see, generally, Bart Hendrickx and Bert Vis, *Energiya-Buran: The Soviet Space Shuttle* (Chichester, UK: Praxis, 2007).

24. Boris Chertok, *The Moon Race,* vol. IV, Rockets and People, ed. Asif Siddiqi (Washington, DC: US Government Printing Office, 2011).

25. Siddiqi, *Challenge to Apollo,* 234.

26. Garber, *Birds of a Feather?* 18.

27. Siddiqi, *Challenge to Apollo*, 835–36.

28. Garber, *Birds of a Feather?* 16.

29. Press-Center, TsAGI, "TsAGI Centenary in the History of Aviation: The Buran Programme," Press Release, Central Aerohydrodynamic Institute, November 15, 2018, http://tsagi.com/pressroom/news/4085/.

30. David S. F. Portree, *Mir Hardware Heritage* (Houston: Information Services Division, Lyndon B. Johnson Space Center, 1995), 63, et seq.; Siddiqi, *Challenge to Apollo*, 591.

31. Portree, *Mir Hardware Heritage*, 63; Siddiqi, *Challenge to Apollo*, 594, 597; Anatoly Zak, "Here Is the Soviet Union's Secret Space Cannon," *Popular Mechanics*, November 16, 2015, https://www.popularmechanics.com/military/weapons/a18187/here-is-the-soviet-unions-secret-space-cannon/.

32. Siddiqi, *Challenge to Apollo*, 837.

33. Garber, *Birds of a Feather?* 19.

34. Alex Roland, "Triumph or Turkey?" *Discover*, 1985.

35. Elaine M. Marconi, "How the Space Shuttle Got 'Smarter,'" NASA.gov, March 24, 2011, https://www.nasa.gov/mission_pages/shuttle/flyout/glass_cockpit.html.

36. Dale D. Myers, "Letter to Robert K. Dawson, Associate Director for Natural Resources, Energy and Science, Office of Management and Budget, January 20, 1988, with Attachment on the Benefits of the Shuttle-C, December 1987," in *Accessing Space*, vol. IV, Exploring the Unknown: Selected Documents in the History of the U.S. Civil Space Program (Washington, DC: NASA History Division, 1999), 395; Marcellus G. Harsh, "Shuttle-C, Evolution to a Heavy Lift Launch Vehicle (Aim-89–2521)" (NASA/Marshall Space Flight Center, 1989).

37. John H. Disher, "Space Transportation: Reflections and Projections," in *Between Sputnik and the Shuttle: New Perspectives on American Astronautics*, ed. Frederick C. Durant, AAS History Series v. 3 (San Diego: Published for American Astronautical Society by Univelt, 1981), 220; Dennis R. Jenkins, *Space Shuttle: Developing an Icon*, Vol. 2 (Forest Lake, MN: Specialty Press, 2016), 457.

38. See, generally, Andrew J. Butrica, *Single Stage to Orbit: Politics, Space Technology, and the Quest for Reusable Rocketry* (Baltimore: Johns Hopkins University Press, 2003).

39. R. Cargill Hall, "Early U. S. Satellite Proposals," *Technology and Culture* 4, no. 4 (1963): 415–416, https://doi.org/10.2307/3101377.

40. Siddiqi, *Challenge to Apollo*.

41. Loyd S. Swenson, Jr., James M. Grimwood, and Charles C. Alexander, *This New Ocean: A History of Project Mercury* (Washington, DC: Scientific and Technical Information Division, Office of Technology Utilization, NASA, 1966), 25.

42. Swenson, Jr. et al., *This New Ocean*, 126.

43. See, for example, Barton C. Hacker and James M. Grimwood, *On the Shoulders of Titans: A History of Project Gemini* (Washington, DC: NASA, 1978);, 140.

44. Butrica, *Single Stage to Orbit*, 85–86.

45. Disher, "Space Transportation: Reflections and Projections," 220.

46. Jenkins, *Space Shuttle: Developing an Icon*, 2: 415.

47. T. A Heppenheimer, *Facing the Heat Barrier: A History of Hypersonics* (Washington, DC: NASA, History Division, 2006), 214.

48. For example, see Department of Defense, Strategic Defense Initiative Organization, "Solicitation for the SSTO Phase II Technology Demonstration, June 5, 1991.," in *Accessing Space*, vol. IV, Exploring the Unknown: Selected Documents in the History of the U.S. Civil Space Program (Washington, DC: NASA History Division, 1999), 584.

49. Butrica, *Single Stage to Orbit*, 7.

50. Heppenheimer, *Facing the Heat Barrier*, 213–218.

51. Heppenheimer, *Facing the Heat Barrier*, 210.

52. Heppenheimer, *Facing the Heat Barrier*, 198.

53. Butrica, *Single Stage to Orbit*, 7.

54. Heppenheimer, *Facing the Heat Barrier*, 222.

55. For example, see Department of Defense, "Report of the Defense Science Board Task Force on the National Aerospace Plane (NASP), September 1988, Pp. 2–25.," in *Accessing Space*, vol. IV, Exploring the Unknown: Selected Documents in the History of the U.S. Civil Space Program (Washington, DC: NASA History Division, 1999), 561–70.

56. Heppenheimer, *Facing the Heat Barrier*, 258.

57. Gary Hudson, "Gary Hudson, Pacific American, Memo to Thomas L. Kessler, General Dynamics/Space Systems Division, 'Comments on SSTO Briefing and a Short History of the Project,' December 17, 1990.," in *Accessing Space*, vol. IV, Exploring the Unknown: Selected Documents in the History of the U.S. Civil Space Program (Washington, DC: NASA History Division, 1999), 575–77.

58. For example, see Department of Defense, Strategic Defense Initiative Organization, "Solicitation for the SSTO Phase II Technology Demonstration, June 5, 1991," 584.

59. For example, see National Aeronautics and Space Administration (NASA), "A Draft Cooperative Agreement Notice—X-33 Phase II: Design and Demonstration, December 14, 1995, Pp. A-2–A-4.," in *Accessing Space*, vol. IV, Exploring the

Unknown: Selected Documents in the History of the U.S. Civil Space Program (Washington, DC: NASA History Division, 1999), 631–634.

60. Heppenheimer, *Facing the Heat Barrier*, 268.

61. The author thanks an anonymous reviewer for this insight.

62. Shayler, *Apollo: The Lost and Forgotten Missions; For All Mankind; Marooned*, John Sturges (dir.) (Los Angeles: Columbia Pictures Corporation, 1969).

63. Robert Truax, "Letter to James C. Fletcher," June 22, 1971, 1, Box 19, Folder 019884, NASA Historical Reference Collection, NASA Headquarters, Washington, DC, 2.

Epilogue

1. Ralph E. Lapp, "Send Computers, Not Men, Into Deep Space," *New York Times*, February 2, 1969, SM32.

2. See, for example, Aristos Georgiou, "Why Did the Space Shuttle Program End?" *Newsweek*, May 21, 2020, https://www.newsweek.com/why-space-shuttle-program-end-1505594.

3. National Aeronautics and Space Administration (NASA), "President Bush Delivers Remarks on U.S. Space Policy," 3–4.

4. John M. Logsdon, *Together in Orbit: The Origins of International Participation in the Space Station* (Washington, DC: NASA History Division, Office of Policy and Plans, NASA Headquarters, 1998).

5. "Mir Space Station," accessed May 10, 2020, https://history.nasa.gov/SP-4225/mir/mir.htm.

6. For example, see David S. F. Portree, *Mir Hardware Heritage* (Houston: Information Services Division, Lyndon B. Johnson Space Center, 1995).

7. See, for example, Matthew H. Hersch, "Using the Shuttle: Operations in Orbit," in *Space Shuttle Legacy: How We Did It and What We Learned*, ed. Roger D. Launius, John Krige, James I. Craig, and Ned Allen, 191–214, Library of Flight Series (Reston, VA: American Institute of Aeronautics and Astronautics, Inc, 2013).

8. "Shuttle-Mir History/References/Documents/Congressional Mir Safety Hearing," accessed March 1, 2023, https://spaceflight.nasa.gov/history/shuttle-mir/references/r-documents-congressional.htm; "Shuttle-Mir History/References/Documents/Administrator's Letter to Congress Concerning Shuttle-Mir Program," accessed March 1, 2023, https://spaceflight.nasa.gov/history/shuttle-mir/references/r-documents-admin-letter.htm; Bryan Burrough, *Dragonfly: NASA and the Crisis aboard the MIR* (New York: HarperCollins, 1998).

9. Roger D. Launius, "The Space Shuttle—Twenty-Five Years On: What Does It Mean to Have Reusable Access to Space?" *Quest* 13 (2006): 10.

10. See, for example, Burrough, *Dragonfly*.

11. Tuong Tran, "Memorandum of Understanding between the National Aeronautics and Space Administration of the United States of America and the Russian Space Agency Concerning Cooperation on the Civil International Space Station," International Space Station (Brian Dunbar, October 23, 2010), https://www.nasa.gov /mission_pages/station/structure/elements/nasa_rsa.html.

12. Georgiou, "Why Did the Space Shuttle Program End?"

13. Dennis R. Jenkins, *Space Shuttle: Developing an Icon, 1972–2013*, vol. 3 (Forest Lake: Specialty Press, 2016), 445.

14. For example, see Kenneth Chang, "On Eve of Retirement, Space Shuttles Capture Museums' Hearts," *New York Times*, March 9, 2011, A1; Dennis R. Jenkins, *Space Shuttle: Developing an Icon*, Vol. 3 (Forest Lake, MN: Specialty Press, 2016), 447.

15. Jenkins, *Space Shuttle: Developing an Icon*, 3: 446.

16. See, generally, Jenkins, *Space Shuttle*, 3:446–479.

17. This material comes from recollections of the author's conversations with NASM staff. See also Valerie Neal, *Spaceflight in the Shuttle Era and Beyond*, 195.

18. For example, see Patrick McGeehan, "Intrepid's Ambitious Mission: To Get Its Own Space Shuttle," *New York Times*, May 14, 2009, A25.

19. Brian Dunbar, "NASA Announces New Homes for Space Shuttle Orbiters after Retirement," NASA (March 24, 2015), http://www.nasa.gov/topics/shuttle_station /features/shuttle_homes.html.

20. For example, see Chang, "On Eve of Retirement, Space Shuttles Capture Museums' Hearts," A3; John Nolan, "Houston Fears Space History Not Enough to Win a Shuttle," *Dayton Daily News*, April 6, 2011, A1; Jenkins, *Space Shuttle: Developing an Icon*, 3: 449–450.

21. Donald K. Slayton and Michael Cassutt, *Deke! U.S. Manned Space: From Mercury to the Shuttle* (New York: St. Martin's Press, 1994), 314.

22. William J. Broad, "Reusable Space 'Truck' For Orbit Experiments," *New York Times*, April 7, 1984, 33.

23. Homer Hickam, "What Makes an Astronaut Crack?" *Los Angeles Times*, February 9, 2007, VCA25.

24. See, generally, Emmanuel Chadeau, *Le rêve et la puissance: L'avion et son siècle* (Paris: Fayard, 1996).

25. See, generally, Matthew H. Hersch, *Inventing the American Astronaut*, Palgrave Studies in the History of Science and Technology (New York: Palgrave Macmillan, 2012).

26. See Amy Paige Kaminski, *Sharing the Shuttle with America: NASA and Public Engagement after Apollo* (PhD diss., Science and Technology in Society, Blacksburg: Virginia Polytechnic Institute, 2015).

27. See, for example, Marina Koren, "Mike Pence's Outer-Space Gospel," *The Atlantic*, August 23, 2018, https://www.theatlantic.com/science/archive/2018/08/mike-pence-nasa-faith-religion/568255/.

28. Edwin T. Layton, *The Revolt of the Engineers: Social Responsibility and the American Engineering Profession* (Cleveland: Case Western Reserve University Press, 1971).

References

2001: A Space Odyssey. Stanley Kubrick (dir.). Los Angeles: Metro-Goldwyn-Mayer, 1968.

Air Force Space Systems Division. "Lunar Expedition Plan: Lunex." US Air Force Systems Command, May 1961. http://www.astronautix.com/data/lunex.pdf.

"All In." *For All Mankind*. Apple TV, June 24, 2022.

Allaway, H. *The Space Shuttle at Work*. Washington, DC: NASA, 1979.

Allen, Joseph P. "Oral History Transcript #3 (Jennifer Ross-Nazzal, Interviewer)." *NASA Johnson Space Center Oral History Project*, Washington, DC, 2004.

Allen, Joseph P. "Oral History Transcript #4 (Jennifer Ross-Nazzal, Interviewer)." *NASA Johnson Space Center Oral History Project*, Washington, DC, 2004.

Arendt, Hannah. *Eichmann in Jerusalem: A Report on the Banality of Evil*. New York: Viking Press, 1963.

Asimov, Isaac. "Letter to James C. Fletcher," August 5, 1976. Ms 202, Box 42, Folder 4. James Chipman Fletcher Papers. Special Collections and Archives. University of Utah, J. Willard Marriott. Salt Lake City.

Armageddon. Michael Bay (dir.). Los Angeles: Touchstone Pictures, 1998.

Atkinson, Joseph D., and Jay M. Shafritz. *The Real Stuff: A History of NASA's Astronaut Recruitment Program*. New York: Praeger, 1985.

Attanasio, Paul. "Movies." *Washington Post*, June 6, 1986, sec. D, 3.

Baldwin, Hanson W. "Dyna-Soar or Dinosaur? Another Military Space Vehicle Due to Become Extinct in Conflict of Ideas View of Air Force Two Forces at Work." *New York Times*, March 16, 1963, 4.

Barnett, Arnold. "Aviation Safety: A Whole New World?" *Transportation Science* 54, no. 1 (January 1, 2020): 84–96.

Bauer, Charles J. "'Rocket' Interceptor Tested." *New York Times*, January 15, 1939, XX8.

BBC News. "The Lyndon Johnson Tapes: Richard Nixon's 'Treason.'" March 22, 2013, sec. Magazine. https://www.bbc.com/news/magazine-21768668.

Beggs, James M. "Letter to Don Fuqua," October 7, 1982. Folder 008960. NASA Historical Reference Collection, NASA Headquarters, Washington, DC.

Berger, Eric. "A Cold War Mystery: Why Did Jimmy Carter Save the Space Shuttle?" *Ars Technica*, July 14, 2016. https://arstechnica.com/science/2016/07/a-cold-war-mystery-why-did-jimmy-carter-save-the-space-shuttle/.

Bilstein, Roger E. *Flight in America, 1900–1983: From the Wrights to the Astronauts.* Baltimore: Johns Hopkins University Press, 1984.

Bilstein, Roger E. *Flight in America: From the Wrights to the Astronauts.* 3rd ed. Baltimore: Johns Hopkins University Press, 2001.

Bilstein, Roger E. *Stages to Saturn: A Technological History of the Apollo/Saturn Launch Vehicles.* Washington, DC: Scientific and Technical Information Branch, NASA, 1980.

Bilstein, R. E., and F. W. Anderson. *Orders of Magnitude: A History of the NACA and NASA, 1915–1990.* Washington, DC: NASA, Office of Management, Scientific and Technical Information Division, 1989.

Bobko, K. J. "Oral History Transcript (Summer Chick Bergen, Interviewer)." *NASA Johnson Space Center Oral History Project*, Houston, 2002.

Boffey, Philip M. "First Shuttle Ride by Private Citizen to Go to Teacher: Hopes of Others Dashed." *New York Times*, August 28, 1984, A1.

Boffey, Philip M. "NASA to Seek Observers to Fly on Shuttles." *New York Times*, December 16, 1983, A11.

Boin, Arjen, and Paul Schulman. "Assessing NASA's Safety Culture: The Limits and Possibilities of High-Reliability Theory." *Public Administration Review* 68, no. 6 (2008): 1050–1062.

Boyne, Walter J. *Beyond the Horizons: The Lockheed Story.* New York: Thomas Dunne Books, 1998.

Brand, Stewart, ed. *The Whole Earth Catalog: Access to Tools.* Menlo Park, CA: Portola Institute, 1968.

Broad, William J. "Reusable Space 'Truck' For Orbit Experiments," *New York Times*, April 7, 1984, 33.

Burgess, Colin. *Selecting the Mercury Seven: The Search for America's First Astronauts.* Chichester, UK: Springer-Praxis, 2011.

Burrough, Bryan. *Dragonfly: NASA and the Crisis Aboard the MIR*. New York: HarperCollins, 1998.

Butrica, Andrew J. *Single Stage to Orbit: Politics, Space Technology, and the Quest for Reusable Rocketry*. Baltimore: Johns Hopkins University Press, 2003.

Cabbage, Michael. *Comm Check . . . : The Final Flight of Shuttle Columbia*. New York: Free Press, 2004.

Campbell, Tom. "Note for Dr. Fletcher Re: Quick Look Summary of Spacelab," August 23, 1976. Ms 202, Box 43, Folder 2. James Chipman Fletcher Papers. Special Collections and Archives. University of Utah, J. Willard Marriott. Salt Lake City.

Campbell-Kelly, Martin, and William Aspray. *Computer: A History of the Information Machine*. Boulder, CO: Westview Press, 2004.

Carpenter, Joel. "PROJECT 1947: German Hypersonic Bomber Prototype?" November 11, 2011. http://www.project1947.com/gfb/antiplofer.htm.

Cassutt, Michael. *The Astronaut Maker: How One Mysterious Engineer Ran Human Spaceflight for a Generation*. Chicago: Review Press, 2018.

Cassutt, Michael. "Secret Space Shuttles." *Air & Space*, 2009. https://www.smithsonianmag.com/air-space-magazine/secret-space-shuttles-35318554/.

Center for the Study of National Reconnaissance Classics. *The CORONA Story*. Chantilly, VA: Center for the Study of National Reconnaissance, 2013.

Ceruzzi, Paul E. "Aerospace Needs, Microelectronics, and the Quest for Reliability: 1962–1975." *Proceedings of the IEEE* 105, no. 7 (July 2017): 1456–1465.

Chadeau, Emmanuel. *Le rêve et la puissance: L'avion et son siècle*. Paris: Fayard, 1996.

Challenger: The Final Flight. Daniel Junge and Steven Leckart (dirs..). Santa Monica, CA, Melbourne, FL, Culver City, CA: Bad Robot, Zipper Bros Films, Sutter Road Picture Company, 2020.

Chang, Kenneth. "On Eve of Retirement, Space Shuttles Capture Museums' Hearts." *New York Times*, March 9, 2011, A1.

Chertok, Boris. *The Moon Race*. Vol. 4. of *Rockets and People*, edited by Asif Siddiqi. Washington, DC: US Government Printing Office, 2011.

Chien, Philip. *Columbia, Final Voyage: The Last Flight of NASA's First Space Shuttle*. New York: Copernicus Books, 2006.

Clark, Evert. "Rocket Plane May Let Astronauts Land at Airfields." *New York Times*, July 12, 1967.

Collins, Harry M., and Trevor J. Pinch. *The Golem at Large: What You Should Know about Technology*. New York: Cambridge University Press, 1998.

Collins, Michael. *Carrying the Fire: An Astronaut's Journeys*. New York: Farrar, 1974.

Columbia Accident Investigation Board (CAIB). *The Columbia Accident Investigation Board Report*. Vol. 1. Washington, DC: Government Printing Office, 2003.

Columbia Accident Investigation Board (CAIB). "STS-107 In-Flight Options Assessment." In *The Columbia Accident Investigation Board Report*. Vol. 2. Washington, DC: Government Printing Office, 2003.

Compton, William David, and Charles D. Benson. *Living and Working in Space: A History of Skylab*. Washington, DC: Scientific and Technical Information Branch, NASA, 1983.

Conquest of Space. Bryan Haskins (dir.). Los Angeles: Paramount Pictures, 1955.

Constant, Edward W. *The Origins of the Turbojet Revolution*. Baltimore: Johns Hopkins University Press, 1980.

Conway, Erik M. "From Rockets to Spacecraft: Making JPL a Place for Planetary Science." *Engineering and Science* 70, no. 4 (2007): 2–10.

Cook, Kevin. *The Burning Blue: The Untold Story of Christa McAuliffe and NASA's Challenger Disaster*. New York: Henry Holt and Company, 2021.

Coopersmith, Jonathan. *Faxed: The Rise and Fall of the Fax Machine*. Baltimore: Johns Hopkins University Press, 2015.

The Core. Jon Amiel (dir.). Los Angeles: Paramount Pictures, 2003.

Corn, Joseph J. *The Winged Gospel: America's Romance with Aviation, 1900–1950*. New York: Oxford University Press, 1983.

Cowan, Ruth Schwartz, and Matthew H. Hersch. *A Social History of American Technology*. 2d ed. New York: Oxford University Press, 2017.

Crippen, Robert L. "Oral History Transcript (Rebecca Wright, Interviewer)." *NASA Johnson Space Center Oral History Project*, 2006.

Cunningham, Walter. *The All-American Boys: An Inside Look at the American Space Program*. Rev. ed. New York: ibooks, 2003.

Cywanowicz, L. "NASA Expands Payload Specialist Opportunities." *NASA News (Marshall Space Flight Center)*. 1982.

Damuhn, M. D. *Back Down to Earth: The Development of Space Policy for NASA during the Jimmy Carter Administration*. PhD diss., University of Wales Swansea, 2000.

Dark Star, John Carpenter (dir.). Swansea, UK: Bryanston Distributing Company, 1974.

David, James E. *Spies and Shuttles: NASA's Secret Relationships with the DOD and CIA*. Gainesville: University Press of Florida, 2015.

Degroot, Gerard J. *Dark Side of the Moon: The Magnificent Madness of the American Lunar Quest.* New York: New York University Press, 2006.

Dean, Margaret Lazarus. "It's the 36th Anniversary of the Space Shuttle Challenger Disaster." *Popular Mechanics*, January 28, 2022. https://www.popularmechanics.com/space/a18616/an-oral-history-of-the-space-shuttle-challenger-disaster/.

"Deep Space Homer." *The Simpsons*. Fox, February 24, 1994.

Delatte, Norbert J. *Beyond Failure: Forensic Case Studies for Civil Engineers*. Reston, VA: ASCE Press, 2009.

Dick, Steven J., ed. *NASA's First 50 Years: Historical Perspectives*. NASA SP 2010–4704. Washington, DC: NASA, Office of Communications, History Division, 2010.

Disher, John H. "Space Transportation: Reflections and Projections." In *Between Sputnik and the Shuttle: New Perspectives on American Astronautics*, edited by Frederick C. Durant, 199–224. AAS History Series v. 3. San Diego: Published for American Astronautical Society by Univelt, 1981.

Donlan, Charles J. "Transmittal of NASA Paper 'Space Shuttle Systems Definition Evolution,' July 11, 1972." In *Exploring the Unknown: Selected Documents in the History of the U.S. Civil Space Program*, IV: 211–214. Washington, DC: NASA History Division, 1999.

Donnelly, J. "Letter to James Fletcher." George M. Low Papers, NASA HQ, Folder 004157, 1973.

Dornberger, Walter R. "The Recoverable, Reusable Space Shuttle." *Aeronautics and Astronautics*, November 1965. Folder 007923. NASA Historical Reference Collection, NASA Headquarters, Washington, DC.

Dornberger, Walter. *V-2*. New York: Viking Press, 1954.

Dunbar, Brian. "NASA Announces New Homes for Space Shuttle Orbiters after Retirement." NASA. March 24, 2015. http://www.nasa.gov/topics/shuttle_station/features/shuttle_homes.html.

Easterbrook, Gregg. "Beam Me Out of This Death Trap, Scotty." *Washington Monthly*, April 1980.

Edgerton, David. *The Shock of the Old: Technology and Global History since 1900*. New York: Oxford University Press, 2007.

Editorial Board. "Bean Counter: O'Keefe Offers Case Study of NASA Don't." *Houston Chronicle*, December 19, 2004. https://www.chron.com/opinion/editorials/article/Bean-counter-O-Keefe-offers-case-study-of-NASA-1970055.php.

Ehricke, Krafft A. "Extraterrestrial Imperative." In *Proceedings*, 139–150. George C. Marshall Space Flight Center: Institute of Navigation, Huntsville, AL, 1971.

Ehricke, Krafft A. "Peenemuende Rocket Center," January 3, 1950. Box 1, Folder 4. Krafft Arnold Ehricke Papers, National Air and Space Museum Archives, Washington, DC.

Ehricke, Krafft A. "Project Orbital Carrier (1st Edition)," May 1952. Box 1, Folder 6. Krafft Arnold Ehricke Papers, National Air and Space Museum Archives, Washington, DC.

Emba, Christine. "Facebook Isn't Too Big to Fail." *Washington Post*, April 10, 2018. https://www.washingtonpost.com/opinions/facebook-isnt-too-big-to-fail/2018/04/10/dd9e9c18-3d02-11e8-a7d1-e4efec6389f0_story.html.

Eskow, Dennis. "Space Vacation 1995." *Popular Mechanics*, 1985, 59–60.

Evans, Michelle. *The X-15 Rocket Plane:* Lincoln: University of Nebraska Press, 2013.

Ezell, Edward C., and Linda N. Ezell. *The Partnership: A History of the Apollo-Soyuz Test Project*. Washington, DC: National Aeronautics and Space Administration, 1978.

Faget, Maxime A. "The Space Shuttle." In *Proceedings*, 7–17. Warrendale, PA: Society of Automotive Engineers, 1980.

Feynman, Richard P. "Appendix F: Personal Observations on the Reliability of the Shuttle." In *Report of the Presidential Commission on the Space Shuttle Challenger Accident*. Washington, DC: Government Printing Office, 1986.

Feynman, Richard P. *"What Do You Care What Other People Think?" Further Adventures of a Curious Character*. New York: W. W. Norton, 2018.

Fleddermann, Charles B. *Engineering Ethics*. 4th ed. Upper Saddle River, NJ: Pearson, 2011.

Fletcher, James C. "Administrator, NASA, to Caspar W. Weinberger, Deputy Director, Office of Management and Budget, December 29, 1971." In *Accessing Space*, IV: 245–249. Exploring the Unknown: Selected Documents in the History of the U.S. Civil Space Program. Washington, DC: NASA History Division, 1999.

Fletcher, James C. "Letter to Elmer B. Staats," February 7, 1977. Ms 202, Box 53, Folder 5. James Chipman Fletcher Papers. Special Collections and Archives. University of Utah, J. Willard Marriott. Salt Lake City.

Fletcher, James C. "Memorandum to Al Lovelace, 'Personal Concern about the Launch Phase of Space Shuttle,' July 7, 1977." In *Accessing Space*, IV: 352–354. Exploring the Unknown: Selected Documents in the History of the U.S. Civil Space Program. Washington, DC: NASA History Division, 1999.

Fletcher, James C. "Memorandum to Associate Administrator for Manned Spaceflight Re: Conversation with Bill Bergen," July 15, 1971. Ms 202, Box 4, Folder 1.

James Chipman Fletcher Papers. Special Collections and Archives. University of Utah, J. Willard Marriott. Salt Lake City.

Fletcher, James C. "Memorandum to Peter M. Flanigan Re: New Names for the Shuttle," December 30, 1971. Ms 202, Box 4, Folder 5. James Chipman Fletcher Papers. Special Collections and Archives. University of Utah, J. Willard Marriott. Salt Lake City.

Fletcher, James C. "Note to Herbert Rowe," August 12, 1976. Ms 202, Box 42, Folder 4. James Chipman Fletcher Papers. Special Collections and Archives. University of Utah, J. Willard Marriott. Salt Lake City.

Fletcher, James C. "Statement by Dr. Fletcher, NASA Administrator." 1972. http://history.nasa.gov/stsnixon.htm.

Fletcher, James C. "The Space Shuttle, November 22, 1971." In *Exploring the Unknown: Selected Documents in the History of the U.S. Civil Space Program*, edited by John M. Logsdon, 555–558. NASA History Series. Washington, DC: National Aeronautics and Space Administration, 1995.

Fletcher, James C., George M. Low, and Richard McCurdy. "Memorandum for the Record, 'Selection of Contractor for Space Shuttle Program,' September 18, 1972." In *Accessing Space*, IV: 262–268. Exploring the Unknown: Selected Documents in the History of the U.S. Civil Space Program. Washington, DC: NASA History Division, 1999.

Flickinger, D. "Biomedical Aspects of Space Flight." In *Man in Space: The United States Air Force Program for Developing the Spacecraft Crew*, edited by K. F. Gantz, 41–63. New York: Duell, 1959.

For All Mankind. Al Reinert (dir.). Los Angeles: Sony Pictures Television, Tall Ship Productions, 2019.

Foster, Amy E. *Integrating Women into the Astronaut Corps: Politics and Logistics at NASA, 1972–2004*. Baltimore: Johns Hopkins University Press, 2011.

Franklin, Ben A. "Miss Woods's Gift List Links Donors to Corporations." *New York Times*, March 20, 1974, 30.

Fritzsche, Peter. *A Nation of Fliers: German Aviation and the Popular Imagination*. Cambridge, MA: Harvard University Press, 1992.

Fullerton, C. G. "Oral History Transcript (Rebecca Wright, Interviewer)." NASA Dryden Flight Research Center: National Aeronautics and Space Administration (NASA), 2002.

Garber, Stephen J. *Birds of a Feather? How Politics and Culture Affected the Designs of the U.S. Space Shuttle and the Soviet Buran*. Master's thesis, Virginia Tech, Blackburg, 2002.

Garber, Stephen J. "Why Does the Space Shuttle Have Wings? A Look at the Social Construction of Technology in Air and Space." *The Flight of STS-1*, April 5, 2001. https://history.nasa.gov/sts1/pages/scot.html.

Garber, Stephen J. "Why Does the Space Shuttle Have Wings? A Look at the Social Construction of Technology in Air and Space (IV. SCOT and the Shuttle's Wings)," July 3, 2014. https://history.nasa.gov/sts1/pages/scotc.html.

Geiger, Clarence J. *History of the X-20A Dyna-Soar*. Vol. 1. Wright-Patterson AFB: Historical Division, Aeronautical Systems Division Information Office, US Air Force, 1963.

Geiger, Clarence J. *Strangled Infant: The Declassified History of the Dyna-Soar Space Plane*. St. Petersburg, FL: Red and Black Publishers, 2019.

Georgiou, Aristos. "Why Did the Space Shuttle Program End?" *Newsweek*, May 21, 2020. https://www.newsweek.com/why-space-shuttle-program-end-1505594.

Geppert, Alexander C. T., ed. *Imagining Outer Space: European Astroculture in the Twentieth Century*. 2nd ed. London: Palgrave Macmillan, 2018.

Geppert, Alexander C. T., ed. *Limiting Outer Space: Astroculture after Apollo*. London: Palgrave Macmillan UK, 2018.

Gilbert, James B. *Redeeming Culture: American Religion in an Age of Science*. Chicago: University of Chicago Press, 1997.

Godwin, Robert. "The Forgotten Plans to Reach the Moon—Before Apollo." *Air & Space Magazine*, July 19, 2019. https://www.airspacemag.com/daily-planet/forgotten-plans-reach-moon-apollo-180972695/.

Goetz, Otto K. "Edited Oral History Transcript (Jennifer Rozz-Nazzal, Interviewer)." *NASA STS Recordation Oral History Project*, 2000.

Goodrich, Malinda K., Alice R. Buchalter, and Patrick M. Miller, eds. *Toward a History of the Space Shuttle: An Annotated Bibliography, Part 2 (1992–2011)*. Washington, DC: NASA History Program Office, 2012.

Government Accounting Office (GAO). "Draft Report to the Congress of the United States, Space Transportation System: Past, Present Future," February 1977. Ms 202, Box 53, Folder 7. James Chipman Fletcher Papers. Special Collections and Archives. University of Utah, J. Willard Marriott. Salt Lake City.

Gravity. Alfonso Cuarón (dir.). Los Angeles: Warner Bros., 2013.

Gray, Edward Z. "Memorandum for the Associate Administrator Re: Saturn S-IC Stage Recovery and Reusability Study," June 8, 1964. Folder 008211. NASA Historical Reference Collection, NASA Headquarters, Washington, DC.

Green, C. M., and M. Lomask. *Vanguard: A History*. Washington, DC: Smithsonian Institution Press, 1971.

Hacker, Barton C. "The Gemini Paraglider: A Failure of Scheduled Innovation, 1961–64." *Social Studies of Science* 22, no. 2 (May 1, 1992): 387–406. https://doi.org/10.1177/030631292022002012.

Hacker, Barton C. "Whoever Heard of Nuclear Ramjets? Project Pluto, 1957–1964." *Journal of the International Committee for the History of Technology* 1 (1995): 85–98.

Hacker, Barton C., and James M. Grimwood. *On the Shoulders of Titans: A History of Project Gemini.* Washington, DC: National Aeronautics and Space Administration, 1978.

Hale, Wayne. "How We Nearly Lost Discovery." *Wayne Hale's Blog,* April 18, 2012. https://waynehale.wordpress.com/2012/04/18/how-we-nearly-lost-discovery/.

Hale, Wayne, ed. *Wings in Orbit: Scientific and Engineering Legacies of the Space Shuttle, 1971–2010.* Washington, DC: National Aeronautics and Space Administration, 2010.

Hall, R. Cargill. "Early U. S. Satellite Proposals." *Technology and Culture* 4, no. 4 (1963): 410–434. https://doi.org/10.2307/3101377.

Harsh, Marcellus G. "Shuttle-C, Evolution to a Heavy Lift Launch Vehicle (Aim-89-2521)." NASA/Marshall Space Flight Center, 1989.

Hartsfield, Henry W., Jr. "Oral History Transcript #2 (Carol Butler, Interviewer)." *NASA Johnson Space Center Oral History Project,* Houston, 2001.

Hartsfield, Henry W., Jr. "Oral History Transcript (Carol Butler, Interviewer)." *NASA Johnson Space Center Oral History Project,* Houston, June 12, 2001.

Harwood, William. "STS-119 Shuttle Report: Legendary Commander Tells Story of Shuttle's Close Call." *Spaceflight Now,* March 27, 2009. https://spaceflightnow.com/shuttle/sts119/090327sts27/.

Headrick, Daniel R. *The Tools of Empire: Technology and European Imperialism in the Nineteenth Century.* New York: Oxford University Press, 1981.

Heaton, Donald H. "Memorandum for the Files," August 23, 1960. Folder 008211. NASA Historical Reference Collection, NASA Headquarters, Washington, DC.

Hecht, Gabrielle. *Being Nuclear: Africans and the Global Uranium Trade.* Cambridge, MA: MIT Press, 2012.

Hecht, Gabrielle, and Michel Callon. *The Radiance of France: Nuclear Power and National Identity after World War II.* Cambridge, MA: MIT Press, 2009.

Heiss, Klaus P., and Oskar Morgenstern. "Memorandum for Dr. James C. Fletcher, Administrator, NASA, 'Factors for a Decision on a New Reusable Space Transportation System,' October 28, 1971." In *Exploring the Unknown: Selected Documents in the*

History of the U.S. Civil Space Program, edited by John M. Logsdon, 449–555. Washington, DC: National Aeronautics and Space Administration, 1995.

Hendrickx, Bart, and Bert Vis. *Energiya-Buran: The Soviet Space Shuttle*. Chichester, UK: Praxis, 2007.

Heppenheimer, T. A. *Development of the Shuttle, 1972–1981*. Washington, DC: Smithsonian Institution Press, 2002.

Heppenheimer, T. A. *Facing the Heat Barrier: A History of Hypersonics*. Washington, DC: National Aeronautics and Space Administration, History Division, 2006.

Heppenheimer, T. A. *The Space Shuttle Decision, 1965–1972*. Washington, DC: Smithsonian Institution Press, 2002.

Hersch, Matthew H. "'Capsules Are Swallowed': The Mythology of the Pilot in American Spaceflight." In *Spacefarers: Images of Astronauts and Cosmonauts in the Heroic Era of Spaceflight*, edited by Michael J. Neufeld., 35–55. Washington, DC: Smithsonian Institution, 2013.

Hersch, Matthew H. *Inventing the American Astronaut*. New York: Palgrave Macmillan, 2012.

Hersch, Matthew H. "Pathfinder to Profit: Lessons from the Space Shuttle Era." NASA Marshall Space Flight Center History Office, Huntsville, AL, 2021.

Hersch, Matthew H. "Redemptive Space: Duty, Death, and the Astronaut-Soldier, 1949–1969." In *We Are All Astronauts: The Image of the Space Traveler in Arts and Media*, edited by Henry Keazor, 35–53. Berlin: Neofelis Verlag, 2019.

Hersch, Matthew H. "Return of the Lost Spaceman: America's Astronauts in Popular Culture, 1959–2006." *The Journal of Popular Culture* 44, no. 1 (February 1, 2011): 73–92. https://doi.org/10.1111/j.1540-5931.2010.00820.x.

Hersch, Matthew H. "Using the Shuttle: Operations in Orbit." In *Space Shuttle Legacy: How We Did It and What We Learned*, edited by Roger D. Launius, John Krige, James I. Craig, and Ned Allen, 191–214. Library of Flight Series. Reston, VA: American Institute of Aeronautics and Astronautics (AIAA), 2013.

Hickam, Homer. "What Makes an Astronaut Crack." *Los Angeles Times*, February 9, 2007, VCA25.

Hill, Gladwin. "X-15 Rocket Plane in Powered Flight." *New York Times*, September 18, 1959, 1.

Hitt, David, and Heather R. Smith. *Bold They Rise: The Space Shuttle Early Years, 1972–1986*. Lincoln: University of Nebraska Press, 2014.

Hohler, Robert T. *I Touch the Future: The Story of Christa McAuliffe*. New York: Random House, 1986.

Hopkins, Andrew. "Was Three Mile Island a 'Normal Accident'?" *Journal of Contingencies and Crisis Management* 9, no. 2 (2002): 65–72. https://doi.org/10.1111/1468-5973.00155.

Houston, Rick. *Wheels Stop: The Tragedies and Triumphs of the Space Shuttle Program, 1986–2011*. Lincoln: University of Nebraska Press, 2013.

Hudson, Gary. "Gary Hudson, Pacific American, Memo to Thomas L. Kessler, General Dynamics/Space Systems Division, 'Comments on SSTO Briefing and a Short History of the Project,' December 17, 1990." In *Accessing Space*, IV: 575–577. Exploring the Unknown: Selected Documents in the History of the U.S. Civil Space Program. Washington, DC: NASA History Division, 1999.

Hughes, Thomas P. *Networks of Power: Electrification in Western Society, 1880–1930*. Baltimore: Johns Hopkins University Press, 1983.

Hughes, Thomas P. "Technological Momentum." In *Does Technology Drive History? The Dilemma of Technological Determinism*, edited by Merritt Roe Smith and Leo Marx, 101–113. Cambridge, MA: MIT Press, 1994.

Hutchinson, Lee. "The Audacious Rescue Plan That Might Have Saved Space Shuttle Columbia." *Ars Technica*, February 1, 2016. https://arstechnica.com/science/2016/02/the-audacious-rescue-plan-that-might-have-saved-space-shuttle-columbia/.

Insurance Journal. "Commercial Air Travel Passenger Death Rate Continues to Decline: MIT Study," January 27, 2020. https://www.insurancejournal.com/news/national/2020/01/27/556562.htm.

Jenkins, Dennis R. *Hypersonics before the Shuttle: A Concise History of the X-15 Research Airplane*. National Aeronautics and Space Administration, NASA Office of Policy and Plans, NASA History Office, NASA Headquarters, 2000.

Jenkins, Dennis R. *Space Shuttle: Developing an Icon, 1972–2013*. Vol. 1. Forest Lake, MN: Specialty Press, 2016.

Jenkins, Dennis R. *Space Shuttle: Developing an Icon, 1972–2013*. Vol. 2. Forest Lake, MN: Specialty Press, 2016.

Jenkins, Dennis R. *Space Shuttle: Developing an Icon, 1972–2013*. Vol. 3. Forest Lake, MN: Specialty Press, 2016.

Jenkins, Dennis R. *Space Shuttle: The History of the National Space Transportation System*. Cape Canaveral, FL: D. R. Jenkins, 2010.

Jenkins, Dennis R., Tony Landis, and Jay Miller. *American X-Vehicles: An Inventory—X-1 to X-50*. Centennial of Flight. Washington, DC: NASA, 2003.

Jenks, Andrew L. *Collaboration in Space and the Search for Peace on Earth*. Anthem Series on Russian, East European and Eurasian Studies. London: Anthem Press, 2022.

Jensen, Claus. *No Downlink: A Dramatic Narrative about the Challenger Accident and Our Time*. New York: Farrar, Straus & Giroux, 1996.

John, Richard R. "Debating New Media: Rewriting Communications History." *Technology and Culture* 64, no. 2 (2023): 308–58.

Jones, Thomas D. *Sky Walking: An Astronaut's Memoir*. New York: Smithsonian Books, 2006.

Jones-Imhotep, Edward. *The Unreliable Nation: Hostile Nature and Technological Failure in the Cold War*. Cambridge, MA: MIT Press, 2017.

Kaminski, Amy Paige. *Sharing the Shuttle with America: NASA and Public Engagement after Apollo*. PhD diss., Science and Technology in Society, Blacksburg: Virginia Polytechnic Institute, 2015.

Karnow, Stanley. *Vietnam: A History*. New York: Viking Press, 1983.

Kevles, Bettyann H. *Almost Heaven: The Story of Women in Space*. New York: Basic Books, 2003.

Koelle, H. H. "Letter to Edward R. Diemer." National Aeronautics and Space Administration, February 8, 1963. Folder 008211. NASA Historical Reference Collection, NASA Headquarters, Washington, DC.

Koren, Marina. "Mike Pence's Outer-Space Gospel." *The Atlantic*, August 23, 2018. https://www.theatlantic.com/science/archive/2018/08/mike-pence-nasa-faith-religion/568255/.

Kraft, Christopher C., Jr. *Flight: My Life in Mission Control*. New York: Dutton, 2001.

Krige, John. *NASA in the World: Fifty Years of International Collaboration in Space*. New York: Palgrave Macmillan, 2013.

Langeswiesche, William. "Should Airplanes Be Flying Themselves?" *Vanity Fair*, September 17, 2014. https://www.vanityfair.com/news/business/2014/10/air-france-flight-447-crash.

Lapp, Ralph E. "Send Computers, Not Men, into Deep Space." *New York Times*, February 2, 1969, SM32.

Launius, Roger D. "A Baker's Dozen of Key Historical Books about the Space Shuttle." *Roger Launius's Blog*, October 19, 2011. https://launiusr.wordpress.com/2011/10/19/a-baker%e2%80%99s-dozen-of-key-historical-books-about-the-space-shuttle/.

Launius, Roger D. "Assessing the Legacy of the Space Shuttle." *Space Policy* 22 (2006): 226–234.

Launius, Roger D. "The Space Shuttle—Twenty-Five Years On: What Does It Mean to Have Reusable Access to Space?" *Quest* 13 (2006): 4–20.

Launius, Roger D. "The Strange Career of the American Spaceplane: The Long History of Wings and Wheels in Human Space Operations." *Centaurus* 55, no. 4 (November 1, 2013): 412–432.

Launius, Roger D. "Summer Reading: Indispensable Books on the History of the Space Shuttle." *Roger Launius's Blog,* June 12, 2017. https://launiusr.wordpress.com/2017/06/12/a-shelf-of-indispensable-books-on-the-space-shuttle/.

Launius, Roger D., and Aaron K. Gillette, eds. *Toward a History of the Space Shuttle: An Annotated Bibliography.* Washington, DC: NASA History Office, 1992.

Launius, Roger D., and Dennis R. Jenkins. *Coming Home: Reentry and Recovery from Space.* Washington, DC: Government Printing Office, 2012.

Launius, Roger D., John Krige, James I. Craig, and Ned Allen (eds.). *Space Shuttle Legacy: How We Did It and What We Learned.* Reston, VA: American Institute of Aeronautics and Astronautics (AIAA), 2013.

Launius, Roger D., and Howard E. McCurdy, eds. "Something Borrowed, Something Blue: Repurposing NASA's Spacecraft." In *NASA Spaceflight: A History of Innovation,* 215–235. New York: Palgrave Macmillan, 2018.

Launius, Roger D., and Howard E. McCurdy, eds. *Spaceflight and the Myth of Presidential Leadership.* Urbana: University of Illinois Press, 1997.

Layton, Edwin T. *The Revolt of the Engineers: Social Responsibility and the American Engineering Profession.* Cleveland: Case Western Reserve University Press, 1971.

Leary, Warren E. "NASA Leader Cites Finances and Submits His Resignation." *New York Times,* December 14, 2004, 20.

Lee, Chester. "Letter to Manager, Space Shuttle Payload Integration and Development Program Office, Johnson Space Center, and Manager, STS Projects Office, Kennedy Space Center, 'Guidelines for Development of the Flight Assignment Baseline,' November 20, 1978." In *Accessing Space,* IV: 344–347. Exploring the Unknown: Selected Documents in the History of the U.S. Civil Space Program. Washington, DC: NASA History Division, 1999.

Lehman, Milton. *This High Man: The Life of Robert H. Goddard.* New York: Farrar, 1963.

Ley, Willy. *The Conquest of Space.* New York: Viking Press, 1949.

Ley, Willy. *Rockets, Missiles, and Space Travel.* Rev. and enlarged ed. New York: Viking Press, 1957.

Lockheed Corporation. "Lockheed Annual Report: 1963." Ann Arbor: ProQuest Annual Reports, 1963. http://search.proquest.com/hnpnewyorktimes/docview/88197575/citation/64159BCECD54CE3PQ/6.

"Lockheed Wins Spaceship Study Pact from NASA." *Boston Globe,* July 10, 1962, 10.

Logsdon, John M. *After Apollo? Richard Nixon and the American Space Program*, London: Palgrave MacMillan, 2015.

Logsdon, John M., ed. *Exploring the Unknown: Selected Documents in the History of the U.S. Civil Space Program*. Vol. 1. Washington, DC: National Aeronautics and Space Administration, 1995.

Logsdon, John M.. *Ronald Reagan and the Space Frontier*. Cham, Switzerland: Springer International Publishing AG, 2018.

Logsdon, John. "The Space Shuttle Program: A Policy Failure?" *Science*, no. 232 (May 30, 1986): 1099–1105.

Logsdon, John M. *Together in Orbit: The Origins of International Participation in the Space Station*. Washington, DC: NASA History Division, Office of Policy and Plans, NASA Headquarters, 1998.

Logsdon, John M.. "Was the Space Shuttle a Mistake?" *MIT Technology Review*, July 6, 2011. https://www.technologyreview.com/s/424586/was-the-space-shuttle-a-mistake/.

Low, George M. "Letter to Donald B. Rice, Assistant Director, Office of Management and Budget, November 22, 1971, with Attached: 'Space Shuttle Configurations.'" In *Accessing Space*, IV: 231–234. Exploring the Unknown: Selected Documents in the History of the U.S. Civil Space Program. Washington, DC: NASA History Division, 1999.

Low, George. "Memorandum for Dale Myers Re: Space Shuttle Objectives," January 27, 1970. Box 19, Folder 019884. NASA Historical Reference Collection, NASA Headquarters, Washington, DC.

Low, George M. "Memorandum to Donald B. Rice, November 22, 1971." In *Accessing Space*, IV: 231–238. Exploring the Unknown: Selected Documents in the History of the U.S. Civil Space Program. Washington, DC: NASA History Division, 1999.

Low, George M. "Memorandum for the Record: 'Space Shuttle Discussions with Secretary Seamans,' January 28, 1970." In *Exploring the Unknown: Selected Documents in the History of the U.S. Civil Space Program*. Vol. 2. Washington, DC: National Aeronautics and Space Administration, 1996.

Low, George M. "Memorandum Re: Luncheon Conversation with Dave Packard," October 20, 1971. Box 19, Folder 019884. NASA Historical Reference Collection, NASA Headquarters, Washington, DC.

Low, George M. "Memorandum Re: Meeting with Sam Phillips," February 28, 1972. Box 19, Folder 019884. NASA Historical Reference Collection, NASA Headquarters, Washington, DC.

Low, George. "Memorandum Re: Space Shuttle Phase B Statement of Work," August 5, 1969. Box 19, Folder 019884. NASA Historical Reference Collection, NASA Headquarters, Washington, DC.

Lucy in the Sky. Noah Hawley (dir.). Los Angeles: Fox Searchlight Pictures, 2019.

Mack, P. E. *Viewing the Earth: The Social Construction of the Landsat Satellite System*. Cambridge, MA: MIT Press, 1990.

Mackowski, M. P. *Testing the Limits: Aviation Medicine and the Origins of Manned Space Flight*. College Station: Texas A&M University Press, 2006.

Maher, Neil M. *Apollo in the Age of Aquarius*. Cambridge, MA: Harvard University Press, 2017.

Malina, Frank J. "Interview with Frank J. Malina." Oral History, 1980. http://resolver.caltech.edu/CaltechOH:OH_Malina_F.

Malina, Frank J. *Notes on Tour of Inspection of the United Kingdom and France*. Pasadena: California Institute of Technology, 1944.

Malina, Frank J. "The Rocket Pioneers." *Engineering & Science*, November 1986, 8–13.

Man in Space. Ward Kimball (dir.). Los Angeles: Walt Disney Productions, 1955.

Marconi, Elaine M. "How the Space Shuttle Got 'Smarter.'" NASA.gov, March 24, 2011. https://www.nasa.gov/mission_pages/shuttle/flyout/glass_cockpit.html.

Mark, Hans. *The Space Station: A Personal Journey*. Durham, NC: Duke University Press, 1987.

Marooned. John Sturges (dir.) Los Angeles: Columbia Pictures Corporation, 1969.

Mathematica. "Economic Analysis of the Space Shuttle System," January 31, 1972. https://ntrs.nasa.gov/api/citations/19730005253/downloads/19730005253.pdf.

McCray, Patrick. *The Visioneers: How a Group of Elite Scientists Pursued Space Colonies, Nanotechnologies, and a Limitless Future*. Course Book. Princeton, NJ: Princeton University Press, 2013.

McCurdy, Howard E. *Inside NASA: High Technology and Organizational Change in the U.S. Space Program*. Baltimore: Johns Hopkins University Press, 1993.

McCurdy, Howard E. *Space and the American Imagination*. Washington, DC: Smithsonian Institution Press, 1997.

McCurdy, Howard E. *The Space Station Decision: Incremental Politics and Technological Choice*, Baltimore: Johns Hopkins University Press, 1990.

McDonald, Allan J. *Truth, Lies, and O-Rings: Inside the Space Shuttle Challenger Disaster*. Gainesville: University Press of Florida, 2009.

McDougall, Walter A. *. . . the Heavens and the Earth: A Political History of the Space Age*. Baltimore: Johns Hopkins University Press, 1997.

McGeehan, Patrick. "Intrepid's Ambitious Mission: To Get Its Own Space Shuttle." *New York Times*, May 14, 2009, sec. New York, A25.

Miles, Marvin. "Advanced X-15 Proposed for Manned Orbital Space Ship: Current Tests Designed for 50-Mile Altitude Because of Heat Problem Advanced X-15 Urged for Manned Space Ship." *Los Angeles Times*, January 11, 1960, B1.

Miles, Marvin. "Wingless 'Lift' Vehicle Nearly Ready for Tests: Northrop Completing Craft That May Be Forerunner of Future Space Shuttle Ship." *Los Angeles Times*, April 23, 1965, 19.

Mindell, David A. *Digital Apollo: Human and Machine in Spaceflight*. Cambridge, MA: MIT Press, 2008.

Mission to Mars. Brian de Palma (dir.). Los Angeles: Buena Vista Distribution, 2000.

Moche, Dinah L. *The Star Wars Question and Answer Book about Space*. New York: Random House, 1979.

Mohun, Arwen. *Risk: Negotiating Safety in American Society*. Baltimore: Johns Hopkins University Press, 2013.

Moonraker. Lewis Gilbert (dir). Los Angeles: United Artists, 1979.

Mork, Brian. "Astronaut Job Interview 2000–2009," 2009. http://www.increa.com/NASA2000/.

Mueller, George E. "Honorary Fellowship Acceptance," address delivered to the British Interplanetary Society, University College London, August 10, 1968, 1–10.

Mueller, George E. "Letter to Kurt H. Debus." National Aeronautics and Space Administration, September 11, 1969. Folder (George Mueller Papers). NASA Historical Reference Collection, NASA Headquarters, Washington, DC.

Mullane, Mike. *Riding Rockets: The Outrageous Tales of a Space Shuttle Astronaut*. New York: Scribner, 2006.

Mumford, Lewis. *Pentagon of Power: The Myth of the Machine,* Vol. 2. New York: Harcourt, Brace Jovanovich, 1974.

Myers, Dale D. "Letter to Robert K. Dawson, Associate Director for Natural Resources, Energy and Science, Office of Management and Budget, January 20, 1988, with Attachment on the Benefits of the Shuttle-C, December 1987." In *Accessing Space*, IV: 394–399. Exploring the Unknown: Selected Documents in the History of the U.S. Civil Space Program. Washington, DC: NASA History Division, 1999.

Nader, R. *Unsafe at Any Speed: The Designed-in Dangers of the American Automobile*. New York: Grossman, 1965.

NASA History Office. "Mir Space Station." Accessed May 10, 2020. https://history.nasa.gov/SP-4225/mir/mir.htm.

National Aeronautics and Space Administration (NASA). "A Draft Cooperative Agreement Notice—X-33 Phase II: Design and Demonstration, December 14, 1995,

A-2–A-4." In *Accessing Space*, IV: 631–634. Exploring the Unknown: Selected Documents in the History of the U.S. Civil Space Program. Washington, DC: NASA History Division, 1999.

National Aeronautics and Space Administration (NASA). "President Bush Delivers Remarks on U.S. Space Policy." *NASA Facts*, January 14, 2004.

National Aeronautics and Space Administration (NASA). *Space Settlements: A Design Study*. SP-413. Washington, DC: National Aeronautics and Space Administration, 1977.

National Aeronautics and Space Administration (NASA). "STS Quarterly Review, Space Shuttle Program," October 27, 1976. Ms 202, Box 43, Folder 3. James Chipman Fletcher Papers. Special Collections and Archives. University of Utah, J. Willard Marriott. Salt Lake City.

National Aeronautics and Space Administration (NASA), Lyndon B. Johnson Space Center. "Major Safety Concerns: Space Shuttle Program, JSC 09990C, November 8, 1976, Preface and pp. 1-1-2-3, 5–1–A-6." In *Accessing Space*, IV: 306–323. Exploring the Unknown: Selected Documents in the History of the U.S. Civil Space Program. Washington, DC: NASA History Division, 1999.

National Aeronautics and Space Administration (NASA), Public Affairs Division. "Our First Quarter Century of Achievement . . . Just the Beginning." Press Release. Washington, DC, October 1983.

National Air and Space Museum. "Lockheed Martin/Boeing RQ-3A DarkStar." Accessed June 5, 2020. https://airandspace.si.edu/collection-objects/lockheed-martin-boeing-rq-3a-darkstar/nasm_A20070230000.

National Air and Space Museum. "Model, Space Shuttle, North American Rockwell Expendable Booster Concept; 1:200," May 1, 2021. https://airandspace.si.edu/collection-objects/model-space-shuttle-north-american-rockwell-expendable-booster-concept-1200/nasm_A19760785000.

National Air and Space Museum. "NASM Oral History Project, Fletcher." Accessed July 23, 2020. https://airandspace.si.edu/research/projects/oral-histories/TRANSCPT/FLETCHER.HTM.

Neal, Valerie. *Spaceflight in the Shuttle Era and Beyond: Redefining Humanity's Purpose in Space*. New Haven, CT, and London: Yale University Press, 2017.

Neufeld, Michael J. *Spacefarers: Images of Astronauts and Cosmonauts in the Heroic Era of Spaceflight*. Washington, DC: Smithsonian Institution, 2013.

Neufeld, Michael J. *Spaceflight: A Concise History*. Cambridge, MA: MIT Press, 2018.

Neufeld, Michael J. "The Von Braun Paradigm and NASA's Long-Term Planning for Human Spaceflight." In *NASA's First 50 Years: Historical Perspectives*, edited by Steven J. Dick, 325–347. Washington, DC: NASA, 2010.

Neufeld, Michael J. *Von Braun: Dreamer of Space, Engineer of War*. New York: A. A. Knopf, 2007.

Neufeld, Michael. "Wernher von Braun, the SS, and Concentration Camp Labor: Questions of Moral, Political, and Criminal Responsibility." *German Studies Review*, 2002, 25 (1): 57–78. https://doi.org/10.2307/1433245.

New York Times. "2 Women Seek Roles as U.S. Space Pilots." July 7, 1963, 15.

New York Times. "6 Pilots Assigned to Dyna-Soar Tests." March 15, 1962, 30.

New York Times. "Dyna-Soar Is Renamed X-20 by the Air Force." June 27, 1962, 5.

New York Times. "Hail Rocket Plane in First Test Hop: Army's New High-Speed Rocket Plane in Test Flight." December 11, 1946, 1.

New York Times. "Kennedy Phones Salute to Pilot: Watches Astronaut On TV." July 22, 1961, 10.

New York Times. "Rocket Plane Downs Japanese Fighter." May 18, 1945, 4.

New York Times. "Rocket Plane Held Army's Prime Need." March 8, 1946, 8.

New York Times. "The Rocket Plane Is Here." January 8, 1944, 12.

New York Times. "Says Pilots like Rocket Plane." January 25, 1944, 4.

New York Times. "Seven Pilots Picked for Satellite Trips: 7 Chosen By U.S. as Space Pilots." April 7, 1959, 1.

New York Times. "Soviet Rocket Plane Foreseen." February 18, 1958, 17.

New York Times. "Space Pilots Get Training on Jets: Future Astronaut Describes Simulated Re-entry." November 10, 1963, 88.

New York Times. "X-15 Test Pilot Decorated as First Winged Astronaut." June 4, 1963, 9.

Newell, Catherine L. *Destined for the Stars: Faith, the Future, and America's Final Frontier*. Pittsburgh: University of Pittsburgh Press, 2019.

"News Release: Space Shuttle Art Exhibition to Go on View at the National Air and Space Museum." Smithsonian Institution, November 12, 1982. Box 1 Folder ("The Artist and the Shuttle: Press"). Smithsonian Institution Traveling Exhibition Service, Exhibition Records, c. 1979–1995, Smithsonian Institution Archives.

"Next: Earth-Orbiting Vacations!" *Christian Science Monitor*. October 19, 1968, Boston, 13.

Nixon, Richard M. "The Statement by President Nixon, 5 January 1972," 1972. http://history.nasa.gov/stsnixon.htm.

Nolan, John. "Houston Fears Space History Not Enough to Win a Shuttle." *Dayton Daily News*, April 6, 2011, A1.

Oberg, James. "Did Pentagon Create Orbital Space Plane?" NBC News Space Analyst Special. MSNBC.com, March 6, 2006. http://www.nbcnews.com/id/11691989/ns/technology_and_science-space/t/did-pentagon-create-orbital-space-plane/.

Office of Management and Budget. "Meeting on the Space Shuttle, November 14, 1979." In *Accessing Space*, IV: 294–305. Exploring the Unknown: Selected Documents in the History of the U.S. Civil Space Program. Washington, DC: NASA History Division, 1999.

Office of Manned Space Flight. "Review of Orbital Transportation Concepts-Low Cost Operations (Integrated Launch/Reentry Vehicle Systems; Round-Trip Operations)." National Aeronautics and Space Administration, December 18, 1968. Folder 008211. NASA Historical Reference Collection, NASA Headquarters, Washington, DC.

Office of the Director of Defense Research & Engineering. "Letter to Robert C. Truax," September 2, 1966. Box 1, Folder 4. Robert C. Truax Collection, National Air and Space Museum Archives, Washington, DC.

O'Leary, Brian. "The Space Shuttle: NASA's White Elephant in the Sky." *Bulletin of the Atomic Scientists* 29, no. 2 (February 1973): 36–43.

Oliver, Kendrick. *To Touch the Face of God: The Sacred, the Profane and the American Space Program, 1957–1975*. Baltimore: Johns Hopkins University Press, 2013.

O'Neill, Gerard K. *High Frontier: Human Colonies in Space*. New York: William Morrow and Company, 1977.

Opinion Research Corporation. "Public Knowledge and Attitudes Regarding Space Exploration," September 1972. Ms 202, Box 10, Folder 6. James Chipman Fletcher Papers. Special Collections and Archives. University of Utah, J. Willard Marriott. Salt Lake City.

O'Toole, Thomas. "Glamor Is Dying in the Camelot of the Astronauts." *Washington Post*, February 5, 1986, A12.

Packard, Vance. *The Waste Makers*. New York: D. McKay Co., 1960.

Paine, Thomas O., and Robert C. Seamans. "Memorandum, April 4, 1969, with Attached 'Terms of Reference for Joint DOD/NASA Study of Space Transportation Systems.'" In *External Relationships*, edited by Dwayne A. Day and John M. Logsdon, Vol. 2. Exploring the Unknown: Selected Documents in the History of the U.S. Civil Space Program. Washington, DC: NASA, 1996.

Paine, Thomas O., and Robert C. Seamans, Jr. "Agreement between the National Aeronautics and Space Administration and the Department of the Air Force Concerning the Space Transportation System," NMI 1052.130, Attachment A, February 17, 1970." In *External Relationships*, edited by Dwayne A. Day and John M. Logsdon, Vol. 2. Exploring the Unknown: Selected Documents in the History of the U.S. Civil Space Program. Washington, DC: NASA, 1996.

Paul, Richard, and Steven Moss. *We Could Not Fail: The First African Americans in the Space Program*. Austin: University of Texas Press, 2015.

Pendás, María González. "Fifty Cents a Foot, 14,500 Buckets: Concrete Numbers and the Illusory Shells of Mexican Economy." *Grey Room* (June 1, 2018): 14–39. https://doi.org/10.1162/grey_a_00240.

Pendle, George. *Strange Angel: The Otherworldly Life of Rocket Scientist John Whiteside Parsons*. New York: Houghton Mifflin Harcourt, 2006.

Perrow, Charles. *Normal Accidents: Living with High-Risk Technologies*. New York: Basic Books, 1984.

Perrow, Charles. *Normal Accidents: Living with High-Risk Technologies: With a New Afterword and a Postscript on the Y2K Problem*. Princeton, NJ: Princeton University Press, 1999.

Peterson, Donald H. "Oral History Transcript (Jennifer Ross-Nazzal, Interviewer)." *NASA Johnson Space Center Oral History Project*, Houston, 2002.

Petty, Chris, and Dennis R. Jenkins. *Beyond Blue Skies: The Rocket Plane Programs That Led to the Space Age*. Illu. ed. Lincoln: University of Nebraska Press, 2020.

Pogue, David. "Escape Refresher 41020, Space Shuttle Crew Escape Briefing Handout, Undated," n.d. NASM-NASM.2006.0013-bx006-fd012_010. David M. Brown Papers, National Air and Space Museum Archives, Washington, DC.

Portree, David S. F. *Humans to Mars: Fifty Years of Mission Planning, 1950–2000*. Washington, DC: NASA History Division, Office of Policy and Plans, 2001.

Portree, David S. F. *Mir Hardware Heritage*. Houston: Information Services Division, Lyndon B. Johnson Space Center, 1995.

Press-Center, TsAGI. "TsAGI Centenary in the History of Aviation: The Buran Programme." *Press Release*. Central Aerohydrodynamic Institute, November 15, 2018. http://tsagi.com/pressroom/news/4085/.

Pynchon, Thomas. *Gravity's Rainbow*. New York: Bantam Books, 1974.

Quora. "Why Did NASA End The Space Shuttle Program?" *Forbes*. Accessed October 21, 2022. https://www.forbes.com/sites/quora/2017/02/02/why-did-nasa-end-the-space-shuttle-program/.

Raudzens, George. "War-Winning Weapons: The Measurement of Technological Determinism in Military History." *The Journal of Military History* 54, no. 4 (1990): 403–434. https://doi.org/10.2307/1986064.

Rauschenberg, Robert. *Hot Shot.* 1983. https://artmuseum.williams.edu/collection/featured-acquisitions/robert-rauschenberg/.

Reagan, Ronald W. "Inaugural Address." Ronald Reagan Presidential Library & Museum, 1981. https://www.reaganlibrary.gov/archives/speech/inaugural-address-1981.

Reed, R. Dale, and Darlene Lister. *Wingless Flight: The Lifting Body Story.* Washington, DC: NASA History Office, Office of Policy and Plans, 1997.

Rhodes, Richard. *The Making of the Atomic Bomb.* New York: Simon & Schuster, 1986.

Ride, Sally K. "Two Small Notebooks Containing Ride's Notes from the Rogers Commission," 1986. NASM-NASM.2014.0025-bx013-fd008_001. Sally K. Ride Papers, National Air and Space Museum Archives, Washington, DC.

Robert Rauschenberg Foundation. "Statement on Hot Shot (1983)," September 4, 2014. https://www.rauschenbergfoundation.org/art/archive/a9.

Robert Rauschenberg Foundation. "Stoned Moon," December 22, 2014. https://www.rauschenbergfoundation.org/art/art-in-context/stoned-moon.

Robles-Anderson, Erica, and Patrik Svensson. "'One Damn Slide after Another': PowerPoint at Every Occasion for Speech." *Computational Culture,* no. 5 (January 15, 2016). http://computationalculture.net/one-damn-slide-after-another-powerpoint-at-every-occasion-for-speech/.

Rocky Jones, Space Ranger. Hollingsworth Morse (dir.). Roland Reed TV Productions, 1954.

Rogers Commission. *Report of the Presidential Commission on the Space Shuttle Challenger Accident.* Vol. 1. Washington, DC: Government Printing Office, 1986.

Roland, Alex. "Triumph or Turkey?" *Discover,* 6, no. 11 (1985): 29–49.

Roland, Alex. "Twin Paradoxes of the Space Age." *Nature* 392, no. 6672 (March 1998): 143–145.

Romick, Darrell. "Concept for a Manned Earth-Satellite Terminal Evolving from Earth-to-Orbit Ferry Rockets," September 17, 1956. Folder 007923. NASA Historical Reference Collection, NASA Headquarters, Washington, DC.

Romick, Darrell C. "METEOR Jr.: A Preliminary Design Investigation of a Minimum Sized Ferry Rocket Vehicle of the METEOR Concept." Goodyear Aircraft Corporation, n.d. Box 2. Darrell C. Romick Papers, National Air and Space Museum Archives, Washington, DC.

Rosenberg, Robert. "Why Shuttle Is Needed, Undated but November 1979." In *Accessing Space,* IV: 292–293. Exploring the Unknown: Selected Documents in the History of the U.S. Civil Space Program. Washington, DC: NASA History Division, 1999.

Rouse, William B. *Failure Management: Malfunctions of Technologies, Organizations, and Society*. Oxford: Oxford University Press, 2021.

Rowe, Herbert J. "Note to Administrator," August 18, 1976. Ms 202, Box 42, Folder 4. James Chipman Fletcher Papers. Special Collections and Archives. University of Utah, J. Willard Marriott. Salt Lake City.

Rumerman, Judy A. *U.S. Human Spaceflight: A Record of Achievement, 1961–2006*. Washington, DC: National Aeronautics and Space Administration, 2007.

Sandage, Scott A. *Born Losers: A History of Failure in America*. Cambridge, MA: Harvard University Press, 2005.

Sänger, E., and J. Bredt. *Tiberelnen Raketenantrleb for Fernboober (A Rocket Drive for Long Range Bombers)*. Translated by M. Hamermesh. Washington, DC: Technical Information Branch, BUAER, Navy Department, 1952.

Sänger, Eugen. *Rocket Flight Engineering*. NASA Technical Translations, F-223. Washington, DC: National Aeronautics and Space Administration, 1965.

Sänger-Bredt, Irene. "The Silver Bird Story." *Spaceflight* 15 (May 1973): 166–181.

Schlager, Neil. *When Technology Fails: Significant Technological Disasters, Accidents, and Failures of the Twentieth Century*. Detroit: Gale Research, 1994.

Schmenck, Harold M., Jr. "A NASA Aide Sees Industry in Space: Near-Perfect Ball Bearings Could Be Made in Orbit." *New York Times*, November 27, 1968, 44.

Schulman, Robert. "NASA Art Program: Space Shuttle Art Collection." National Aeronautics and Space Administration, n.d. Box 1 Folder ("The Artist and the Shuttle: Organizer"). Smithsonian Institution Traveling Exhibition Service, Exhibition Records, c. 1979–1995, Smithsonian Institution Archives.

Scott, David Meerman, and Richard Jurek. *Marketing the Moon: The Selling of the Apollo Lunar Program*. Cambridge, MA: MIT Press, 2014.

Seamans, Robert C., Jr. "Letter to Honorable Spiro T. Agnew, Vice President, August 4, 1969." In *Exploring the Unknown: Selected Documents in the History of the U.S. Civil Space Program*, edited by John M. Logsdon, 519–522. NASA History Series. Washington, DC: National Aeronautics and Space Administration, 1995.

Seamans, Robert C. Jr., and Frederick I. Ordway, III. "Lessons of Apollo for Large-Scale Technology." In *Between Sputnik and the Shuttle: New Perspectives on American Astronautics*, edited by Frederick C. Durant, 241–288. AAS History Series v. 3. San Diego: Published for American Astronautical Society by Univelt, 1981.

Serber, Robert, and Richard Rhodes. *The Los Alamos Primer: The First Lectures on How to Build an Atomic Bomb, Updated with a New Introduction by Richard Rhodes*. Berkeley: University of California Press, 2020.

Shayler, David J. *Apollo: The Lost and Forgotten Missions*. Chichester, UK: Springer, 2002.

Shayler, David J. *NASA's Scientist-Astronauts*. New York: Springer, 2007.

Shayler, David J. *Skylab: America's Space Station*. Chichester, UK: Praxis, 2001.

Sheldon, Courtney. "Planet Shuttle Proposed: Steps Outlined." *Christian Science Monitor*. December 9, 1959.

Sherry, Michael S. *The Rise of American Air Power the Creation of Armageddon*. Philadelphia: University of Pennsylvania Press, 2006.

"Shuttle-Mir History/References/Documents/Administrator's Letter to Congress Concerning Shuttle-Mir Program." August 29, 1997. Accessed March 1, 2023. https://historycollection.jsc.nasa.gov/history/shuttle-mir/references/to-r-documents-admin-letter.htm.

"Shuttle-Mir History/References/Documents/Congressional Mir Safety Hearing." Sept. 18, 1997. Accessed March 1, 2023. https://historycollection.jsc.nasa.gov/history/shuttle-mir/references/r-documents-congressional.htm.

Siddiqi, Asif A. *Challenge to Apollo: The Soviet Union and the Space Race, 1945–1974*. Washington, DC: National Aeronautics and Space Administration, NASA History Division, Office of Policy and Plans, 2000.

Smithsonian Institution Traveling Exhibition Service. "The Artist and the Space Shuttle: Itinerary." Smithsonian Institution, November 12, 1982. Box 1 Folder ("The Artist and the Shuttle: Press"). Smithsonian Institution Traveling Exhibition Service, Exhibition Records, c. 1979–1995, Smithsonian Institution Archives.

"Skylab B," 2019. http://www.astronautix.com/s/skylabb.html.

Slayton, Donald K., and Michael Cassutt. *Deke! U.S. Manned Space: From Mercury to the Shuttle*. New York: St. Martin's Press, 1994.

Smith, R. Jeffrey. "Estrangement on the Launch Pad: DOD Loses Affection for the Space Shuttle and Takes up with an Old Flame." *Science* 224 (4656): 1407–1409. https://doi.org/10.1126/science.224.4656.1407.

Smith, Robert W. *The Space Telescope: A Study of NASA, Science, Technology, and Politics*. New York: Cambridge University Press, 1989.

Smithsonian Institution. "The Artist and the Space Shuttle," December 1982, 74–81.

Society of Experimental Test Pilots, ed. "Testy Test Pilots Society (For Lack of a Better Name at the This Point): Minutes of the First Organized Meeting." In *History of the First 20 Years*, 10. Covina: Taylor Pub. Co., 1978.

SpaceCamp. Harry Winer (dir.). Los Angeles: Twentieth Century Fox, 1986.

Space Cowboys. Clint Eastwood (dir.). Los Angeles: Warner Bros., 2000.

Space Is the Place. John Coney (dir.). North American Star System, 1974.

Space Task Group, National Aeronautics and Space Administration (NASA). "Handwritten Revisions to 12/30/1960 Telex from Loudon Wainwright, Life Magazine," 1960. Records of the National Aeronautics and Space Administration. National Archives and Records Center at College Park, MD.

Space Task Group, National Aeronautics and Space Administration (NASA). "NASA Project A, Announcement 1," December 22, 1958. Folder 013880. NASA Historical Reference Collection, NASA Headquarters, Washington, DC.

Stangneth, Bettina. *Eichmann before Jerusalem: The Unexamined Life of a Mass Murderer*. New York: Knopf Doubleday Publishing Group, 2014.

Starflight: The Plane That Couldn't Land. ABC, February 27, 1983.

Steer, Cassandra, and Matthew H. Hersch (eds.). *War and Peace in Outer Space: Law, Policy, and Ethics*. Oxford: Oxford University Press, 2021.

Stever, H. Guyford. "Letter from Panel on Redesign of Space Shuttle Solid Rocket Booster, Committee on NASA Scientific and Technological Program Reviews, National Research Council, to James C. Fletcher, Administrator, NASA, Seventh Interim Report, September 9, 1986." In *Exploring the Unknown: Selected Documents in the History of the U.S. Civil Space Program*, IV: 385–393. Washington, DC: NASA History Division, 1999.

Swenson, Jr., Loyd S., James M. Grimwood, and Charles C. Alexander. *This New Ocean: A History of Project Mercury*. Washington, DC: Scientific and Technical Information Division, Office of Technology Utilization, National Aeronautics and Space Administration, 1966.

Taraborrelli, J. Randy. *The Hiltons: The True Story of an American Dynasty*. New York: Grand Central Publishing, 2014.

Thomas, Kenneth S., and Harold J. McMann. *U.S. Spacesuits*. Chichester, UK: Praxis, 2006.

Thompson, William C., and United States. *Dynamic Model Investigation of the Landing Characteristics of a Manned Spacecraft*. NASA TN D-2497. Washington, DC: NASA Langley Research Center, 1965.

Tompkins, Phillip K., and Emily V. Tompkins. *Apollo, Challenger, Columbia: The Decline of the Space Program: A Study in Organizational Communication*. Los Angeles: Roxbury Publishing, 2005.

Townes, Charles, et al., "Report of the Task Force on Space, January 8, 1969." In *Exploring the Unknown: Selected Documents in the History of the U.S. Civil Space Program*, edited by John M. Logsdon, 499–512. NASA History Series. Washington, DC: National Aeronautics and Space Administration, 1995.

Tran, Tuong. "Memorandum of Understanding Between the National Aeronautics and Space Administration of the United States of America and the Russian Space Agency Concerning Cooperation on the Civil International Space Station." International Space Station. October 23, 2010. https://www.nasa.gov/mission_pages/station/structure/elements/nasa_rsa.html.

Tribbe, Matthew D. *No Requiem for the Space Age: The Apollo Moon Landings and American Culture*. Oxford: Oxford University Press, 2014.

Truax, Robert C. "From Airlines to Spacelines?" 1963. Box 18, Folder 11. Robert C. Truax Collection, National Air and Space Museum Archives, Washington, DC.

Truax, Robert. "Letter to James C. Fletcher," June 22, 1971. Box 19, Folder 019884. NASA Historical Reference Collection, NASA Headquarters, Washington, DC.

Truax, Robert C. "The Pressure-Fed Booster–Dark Horse of the Space Race." XIXth International Astronautical Congress, 1968. Box 18, Folder 12. Robert C. Truax Collection, National Air and Space Museum Archives, Washington, DC.

Truax, Robert C. "Shuttles—What Price Elegance?" *Astronautics and Aeronautics* 8 (June 1970), 22–23.

Tufte, Edward R. *The Cognitive Style of PowerPoint*. Cheshire: Graphics Press, 2003.

Ulrich, Bert. "NASA and the Arts." NASA.gov. July 3, 2013. https://www.nasa.gov/50th/50th_magazine/arts.html.

US Department of Defense. "Report of the Defense Science Board Task Force on the National Aerospace Plane (NASP), September 1988, pp. 2–25." In *Accessing Space*, IV: 561–570. Exploring the Unknown: Selected Documents in the History of the U.S. Civil Space Program. Washington, DC: NASA History Division, 1999.

US Department of Defense, Strategic Defense Initiative Organization. "Solicitation for the SSTO Phase II Technology Demonstration, June 5, 1991." In *Accessing Space*, IV: 577–584. Exploring the Unknown: Selected Documents in the History of the U.S. Civil Space Program. Washington, DC: NASA History Division, 1999.

US House Committee on Science and Technology. "Investigation of the Challenger Accident." Washington, DC: Government Printing Office, October 29, 1986. https://www.govinfo.gov/content/pkg/GPO-CRPT-99hrpt1016/pdf/GPO-CRPT-99hrpt1016.pdf.

Van Hooser, Katherine P. "Space Shuttle Main Engine—the Relentless Pursuit of Improvement." In *NASA Center for AeroSpace Information (CASI), Conference Proceedings*. Hampton, VA: NASA/Langley Research Center, 2012.

Vaughan, Diane. *The Challenger Launch Decision: Risky Technology, Culture, and Deviance at NASA*. Chicago: University of Chicago Press, 1996.

Vinsel, Lee, and Andrew L. Russell. *The Innovation Delusion: How Our Obsession with the New Has Disrupted the Work That Matters Most*. New York: Currency, 2020.

von Braun, Werhner. "Can We Get to Mars?" *Collier's*, 1954, 22–29.

von Braun, Werhner. *The Mars Project*. Urbana: University of Illinois Press, 1953.

von Braun, Wernher, F. I. Ordway, and D. Dooling. *Space Travel: A History*. 4th ed. New York: Harper & Row, 1985.

von Braun, Wernher, and Cornelius Ryan. *Conquest of the Moon*. New York: Viking Press, 1953.

Wade, Mark. "Apollo L-2C." *Encyclopedia Astronautica*, 2019. http://www.astronautix.com/a/apollol-2c.html.

Wade, Mark. "Saenger." *Encyclopedia Astronautica*, March 30, 2019. http://www.astronautix.com/s/saenger.html.

Wall Street Journal. "Lockheed Wins Contract to Design a Rocket Ship for a Space Shuttle." July 10, 1962, 10.

Weber, Arnold R. "Memorandum for Peter Flanigan, 'Space Shuttle Program,' June 10, 1971, with Attached: 'NASA's Internal Organization for the Space Shuttle Project' and 'NASA's Space Shuttle Program.'" In *Accessing Space*, IV:249–52. Exploring the Unknown: Selected Documents in the History of the U.S. Civil Space Program. Washington, DC: NASA History Division, 1999.

Weinberger, Caspar W., and George P. Shultz. "Memorandum for the President, 'Future of NASA,' August 12, 1971." In *Exploring the Unknown: Selected Documents in the History of the U.S. Civil Space Program*, edited by John M. Logsdon, 546. Washington, DC: NASA History Office, 1995.

Weitekamp, Margaret A. *Space Craze: America's Enduring Fascination with Real and Imagined Spaceflight*. Washington, DC: Smithsonian Books, 2022.

Wellerstein, Alex. "The Demon Core." *The New Yorker*, May 21, 2016. https://www.newyorker.com/tech/annals-of-technology/demon-core-the-strange-death-of-louis-slotin.

West, Julian G. "The Atomic-Bomb Core That Escaped World War II." *The Atlantic*, https://www.theatlantic.com/technology/archive/2018/04/tickling-the-dragons-tail-plutonium-time-bomb/557006/.

Westwick, Peter J. *Into the Black: JPL and the American Space Program, 1976–2004*. New Haven, CT: Yale University Press, 2011.

White, Rowland. *Into the Black: The Extraordinary Untold Story of the First Flight of the Space Shuttle Columbia and the Men Who Flew Her*. New York: Touchstone, 2016.

Wilford, John Noble. "A Teacher Trains for Outer Space." *New York Times*, January 5, 1986, SM16.

Williamson, Ray A. "Developing the Space Shuttle: Early Concepts of a Reusable Launch Vehicle." In *Accessing Space*, IV:161–93. Exploring the Unknown: Selected Documents in the History of the U.S. Civil Space Program. Washington, DC: NASA History Division, 1999.

Wilson, Tim. "External Tank Tiger Team Report," 2005, 80. https://www.nasa.gov/pdf/136149main_ET_tiger_team_report.pdf.

Winston, J. *The Making of the Trek Conventions: Or, How to Throw a Party for 12,000 of Your Most Intimate Friends*. New York: Doubleday, 1977.

Witkin, Richard. "2 Teams Bidding on Space Glider: Air Force Likely to Narrow Dyna-Soar Race to Boeing and Bell-Martin Groups." *New York Times*, May 11, 1958, 41.

Witkin, Richard. "X-15 Rocket Plane Is Unveiled by U.S." *New York Times*, October 16, 1958, 1.

Wolfe, Tom. *The Right Stuff*. New York: Farrar, Straus, and Giroux, 1979.

Yardley, John F. "Letter to Director, Public Affairs, NASA, Memorandum, 'Recommended Orbiter Names,' May 26, 1978, with Attached: 'Recommended List of Orbiter Names.'" In *Accessing Space*, IV: 274–275. Exploring the Unknown: Selected Documents in the History of the U.S. Civil Space Program. Washington, DC: NASA History Division, 1999.

Yardley, John F. "Study of TPS Inspection and Repair On-Orbit, June 14, 1979." In *Accessing Space*, IV: 282–283. Exploring the Unknown: Selected Documents in the History of the U.S. Civil Space Program. Washington, DC: NASA History Division, 1999.

Young, John W. "One Part of the 51-L Accident—Space Shuttle Program Flight Safety, March 4, 1986, with Attached: 'Examples of Uncertain Operational and Engineering Conditions or Events Which We "Routinely" Accept Now in the Space Shuttle Program.'" In *Exploring the Unknown: Selected Documents in the History of the U.S. Civil Space Program*, IV: 378–382. Washington, DC: NASA History Division, 1999.

Zak, Anatoly. "Here Is the Soviet Union's Secret Space Cannon." *Popular Mechanics*, November 16, 2015. https://www.popularmechanics.com/military/weapons/a18187/here-is-the-soviet-unions-secret-space-cannon/.

Zucrow, Maurice J. *Principles of Jet Propulsion and Gas Turbines*. New York: John Wiley, 1948.

Index

Page numbers followed by *t* indicate tables and *f* indicate figures.